European Multinationals
in Core Technologies

WILEY/IRM SERIES ON MULTINATIONALS

Appropriate or Underdeveloped Technology?
Arghiri Emmanuel

The Threat of Japanese Multinationals: Can the West Respond?
Lawrence G. Franko

The New Multinationals: The Spread of Third World Enterprises
Sanjaya Lall

International Disclosure and the Multinational Corporation
S. J. Gray, J. C. Shaw, L. B. McSweeney

Multinational Enterprises, Economic Structure and International
Competitiveness
Edited by John Dunning

Britain and the Multinationals
John Stopford and Louis Turner

State-Owned Multinationals
J. P. Anastassopoulos, G. Blanc, P. Dussauge

European Multinationals in Core Technologies
Rob van Tulder and Gerd Junne

European Multinationals in Core Technologies

ROB VAN TULDER and GERD JUNNE

Department of International Relations
University of Amsterdam

JOHN WILEY & SONS
Chichester · New York · Brisbane · Toronto · Singapore

Copyright ©1988 by I.R.M., Geneva

Library of Congress Cataloging-in-Publication Data:
Tulder, Rob van.
  European multinationals in core technologies.

  (Wiley/IRM series on multinationals)
  Bibliography: p.
  Includes index.
  1. High technology industries—Europe. 2. Inter-
national business enterprises—Europe. 3. Technological
innovations—Europe. 4. Technology and state—Europe.
I. Junne, Gerd. II. Title. III. Series.
HC240.9.H53T86   1988   338.8'87   87-25398
ISBN 0 471 91802 4

British Library Cataloguing in Publication Data:
Tulder, Rob van
  European multinationals in core technologies.
  —(Wiley–IRM series on multinationals).
  1. International business enterprises—
  Europe 2. Technological innovations—
  Europe
  I. Title II. Junne, Gerd
  338.8'8894   HD2844

ISBN 0 471 91802 4

Phototypeset by Dobbie Typesetting Service, Plymouth, Devon
Printed in Great Britain by St. Edmundsbury Press Ltd, Bury St. Edmunds

Contents

PREFACE viii

INTRODUCTION ix

CHAPTER ONE — **The Challenge of New Technologies**   1

1.1. Obstacles to further accumulation   2
1.2. The emphasis on 'high technology'   5
    1.2.1. Microelectronics: the present revolution   7
    1.2.2. The next revolution around the corner: biotechnology 12
    1.2.3. Increased convergence between and within the clusters 15
1.3. Coping with accumulation problems 19
    1.3.1. Core technologies save labour 20
    1.3.2. Core technologies save capital 21
    1.3.3. Core technologies save raw materials 22
    1.3.4. Core technologies reduce energy consumption 23
    1.3.5. Core technologies help to fight pollution 24
    1.3.6. Core technologies provide more flexibility 26

CHAPTER TWO — **Corporate Strategies Towards Core**
                   Technologies 28

2.1. Semiconductors: the choice between standard and custom 32
2.2. Computers: the continuous battle with IBM 40
2.3. Telecommunications equipment: merging to stay in the race 44
2.4. Robotics: target area for machine tool, automotive and electronics
    companies 46
2.5. Computer Aided Design (CAD): key to further progress 50
2.6. Software: priority for captive production 53
2.7. Biotechnology: target area for pharmaceutical, (petro)chemical
    and food-processing companies 56
2.8. New materials 61
2.9. The increasing importance of R&D for core technologies 63
2.10. A division of labour between European multinationals 70

CHAPTER THREE — **Changing Structure of Multinationals** . . . 74

3.1. Impact of new technologies on the structure of car
    manufacturers 76
3.2. Internal reorganization 82
    3.2.1. The consequences of converging technologies 82
    3.2.2. The creation of Independent Business Units 84
    3.2.3. Centralization of automation activities 85

3.2.4. The restructuring of the research and development organization   88
3.2.5. Centralization of communication and decision-making
       structure   90
3.3. Institutionalization of a subcontractor structure   92
3.4. Increased capital intensity and declining direct labour content ...   94

CHAPTER FOUR—Social Impact of New Technologies   97

4.1. The impact on employment...................................   98
    4.1.1. The 'Meta-Studie' of the German Ministry of Research and
           Technology   98
    4.1.2. Direct versus indirect effects   100
    4.1.3. The contradictory impact of product and process innovation   101
    4.1.4. Microeconomic versus macroeconomic perspective........   106
    4.1.5. Distinction between large and small firms   107
    4.1.6. Short versus long term   108
    4.1.7. Conclusion   111
4.2. Multinationals, new technologies and qualification   112
4.3. Impact on regional investment patterns   116
4.4. Impact on workers' participation and trade union influence   120
4.5. Conclusion   123

**CHAPTER FIVE—International Competitiveness of European
                High Tech Firms**   125

5.1. Special features of European multinationals   126
    5.1.1. Smaller size: a problem of the past   126
    5.1.2. Consequences of higher degree of diversification   126
    5.1.3. Determinants of competition: market dominance versus
           technology dominance   129
5.2. Competitiveness in four core technologies   130
    5.2.1. Competitiveness in semiconductors   130
    5.2.2. Competitiveness in telecommunications   133
    5.2.3. Manufacturing technology.............................   138
    5.2.4. Competitive position in biotechnology   141
5.3. The social setting   146
    5.3.1. The university–industry relationship   147
    5.3.2. The cost of capital   148
    5.3.3. Human resources.....................................   150
    5.3.4. The quality of management   152
5.4. Conclusion   155

**CHAPTER SIX—High Tech Multinationals and National
               Innovation Policies**.......................   156

6.1. The international subsidy race   158
    6.1.1. Government intervention before the 1970s..............   159
    6.1.2. Stimulation of innovation as a response to the economic crisis   162

6.1.3. Institutional changes.................................. 163
6.1.4. Government programmes in microelectronics 166
6.1.5. Subsidies for biotechnology 172
6.2. Governments and large companies: a changing balance of power? 177
    6.2.1. Concentration of national R&D on large companies 177
    6.2.2. European multinationals as prime defence contractors 185
    6.2.3. The dominance of national champions in the
        telecommunications industry......................... 188
    6.2.4. A shift in the balance of power between government
        and companies...................................... 191
6.3. The regulatory framework.................................. 197
    6.3.1. Anti-antitrust measures 198
    6.3.2. Controlled deregulation 201
6.4. Conclusion 206

CHAPTER SEVEN — **International Cooperation in
                  Core Technologies**.................... 209

7.1. Europeanization of government support..................... 210
    7.1.1. Overlapping national efforts......................... 210
    7.1.2. Counterproductive effects of national efforts 210
    7.1.3. The beginning of a European technology policy.......... 211
    7.1.4. The commitment of European multinationals to a
        European approach................................. 213
7.2. Reasons for international cooperation 217
    7.2.1. Risk sharing 217
    7.2.2. Market penetration in spite of protectionism 218
    7.2.3. Many-fold applications of core technologies 218
    7.2.4. Complementary developments in different technology clusters 218
    7.2.5. The need for systems instead of stand-alone products..... 219
    7.2.6. Establishment of common standards.................... 220
    7.2.7. The dependence of competitiveness on the right
        cooperation agreements 220
7.3. European networks of cooperation 222
    7.3.1. The European programmes 222
    7.3.2. The dominance of large companies in European programmes 225
    7.3.3. Patterns of cooperation 231
7.4. Cooperation with American and Japanese companies 233
    7.4.1. The pattern of cooperation with American and Japanese firms 234
    7.4.2. European cooperation: no substitute for intercontinental
        cooperation.. 247
7.5. Conclusion 250

CONCLUSION... 253
BIBLIOGRAPHY 260
INDEX... 271

Preface

In 1983 the Institute for Research and Information on Multinationals (IRM) approached us to prepare a number of pilot studies on 'The Role of Multinationals in European Technology Production'. The four resulting studies on robotics, telecommunications, microprocessors and bio-technology described the activities of European multinationals in these sectors, assessed their international competitiveness, and discussed the interaction with national governments and the pattern of international cooperation within Europe and between European and American and Japanese companies.

The pilot studies were finished in 1984. The response to the four studies made clear that a more elaborate and integrated treatment of the role and activities of European multinationals in high technology fields would be useful. The present volume is the result of this renewed effort. In the process of writing the book, we were supported by Michiel Roscam Abbing, who kept the information up to date and assisted in many other areas, and by Paul Wiltgen who prepared an in-depth analysis of the research and development efforts of selected European multinationals. We thank Jeffrey Hart for his very thorough comments on a previous version of the manuscript. We could also draw on the results of a parallel research project funded by the Netherlands Organization for the Advancement of Pure Research (ZWO) on the international restructuring race, carried out by Guido Ruivenkamp and Rob van Tulder, and on the results of research on the development and diffusion of new technologies by Annemieke Roobeek.

We wish to thank Eneko Landaburu, director of IRM during the period that this book was written, and Geoffrey Hamilton, head of research at IRM, for their very stimulating and helpful cooperation in the preparation of this book. We deplore that the IRM had to close right at the moment that it had achieved its objective to become an internationally respected institution for research and information on multinationals. This has not been an isolated phenomenon but represents a general decline in attention to the study of multinational corporations and especially the role they play in society at large. One of the results of this book in any case is that a decline of interest in the study of multinationals is not justified. On the contrary.

July 1987
Amsterdam

ROB VAN TULDER
GERD JUNNE

Introduction

The depth of the economic recession of the 1970s and the early 1980s and the slow pace of recovery have underlined that the problems of the world economy were not only due to a downturn in the business cycle. More structural changes were under way. Comparisons with former deep structural crises have shown that it has often been the application of a *new set of technologies* that has created the precondition for a new general upswing. This insight has intensified the interest in 'new technologies'. New technologies such as *microelectronics, biotechnology or new materials* are expected to give rise to new products and to new production processes which will provide new avenues for economic growth. They will also profoundly change our lives. They will affect our consumption pattern, the labour process, company structures, the destiny of national industries as well as the patterns of international competition. Their importance has been recognized by managers and politicians. Companies have increased their research and development (R&D) efforts and try to adapt their product range and their internal structure to the exigencies of this development. Governments boost their support for the development and diffusion of new technologies. The future impact of these new technologies is at the centre of public attention.

Technological change does not take place in a vacuum. It is to a large extent given shape by private companies who invest in R&D in order to realize an extra profit. There has been an intensive debate on whether small or large companies are more innovative, in which three major positions have come forward. The first is that large firms tend to rest on their monopolistic laurels and tend to become less innovative, thus providing continuous new opportunities for small innovative firms. The second is that there is no association between firm size and relative scale of research and inventive activities. According to the third, the largest firms are not only the most R&D intensive, but are also responsible for most of the innovations, especially when it comes to the introduction of 'innovative systems' in which a number of inventions from different fields are *combined*. Whereas the debate may be open as far as developments in the United States are concerned, the situation is clearly different in Western Europe, where multinational corporations are undoubtedly the most important 'carriers' of new technologies. Whereas small companies certainly make important contributions to the innovation process, the large companies are mainly responsible for the commercialization of these innovations. Even for the technologies developed outside the corporate sector, their selection often decides which technologies are applied in new products and processes and which are not.

Multinationals, certainly in Europe, have become crucial for technological development. But technological development has also become crucial for the multinationals: to stay competitive, they have to master and to apply the new technologies as early as possible. Not all new technologies, however, are equally important. In this book, we introduce the concept of *core technologies* which are of a specific strategic value to companies as well as to countries. Core technologies are not only those new technologies that are applicable in many products and processes in different industries and thus form the basis of an industrial renewal of many sectors of the economy. They are also the technologies that at the same time promise to make a contribution to overcome those problems that have been created by the fast accumulation of the postwar period. The concept of *core technologies* is further elaborated in Chapter One of this book. It also discusses the convergence of the technologies that will give shape to the 'Factory-of-the-future', the 'Office-of-the-future' and the 'Farm-of-the-future'.

It is our thesis that practically all major corporations are shifting their activities into the direction of these core technologies. This shift can explain the sometimes confusing pattern of specialization and diversification of large companies. We expect that those companies that have already a strong foothold in core technologies will *specialize* in these fields and hardly diversify into other areas. They may even give up some of their earlier acquisitions in other fields in order to concentrate on their main business. Where they do diversify, they tend to follow 'technological trajectories': they become engaged in fields where they can apply their technological advantage. Companies, on the other hand, which have less experience with core technologies, will diversify into core technologies. They will either try to build up some in-house expertise in these fields, or acquire (a share in) companies which are already strong in this area. There are certainly additional reasons to explain the diversification and specialization patterns of any given company, but the shift towards core technologies seems to be a major trend in the overall pattern of the corporate sector as a whole.

In Chapter Two we shall see how far we find this trend in the strategy of major European multinational corporations.* We analyse the activity of about 40 European multinationals in core technologies. The activity of these corporations is used as an empirical illustration for our argument throughout this book. This group comprises most of the largest European

*A selection of firms will appear in capitals throughout the text. The selection criteria are mentioned in note 82 of Chapter Two.

multinationals. Some smaller multinational companies have been included that contribute in an important way to European activity in core technologies. We have not confined ourselves to the analysis of these selected companies, however, but have included evidence on other companies whenever it has added relevant information.

An expansion of activity in the field of core technologies has far-reaching effects, both inside a company, for a company's immediate environment, and for the pattern of international competition. These impacts will be the subject of consecutive chapters in this book.

Multinationals do not only develop, produce and sell technology and technology intensive products. Multinationals themselves are the *most important appliers of new technologies* as well. Our thesis is that the application of new technologies will change multinational corporations in a profound way. As a result, *even the very concept of what a multinational corporation is may have to change.* The development of new design, production and communication equipment makes it possible that one single headquarters in one country can direct and control far-flung activities in many countries around the globe without owning any of the productive units any longer—eventually even without possessing any formal assets abroad. Such a company actually would not fall under any definition of a multinational or transnational corporation as these terms are used today. Still, it could be a company that *organizes* international activities on the same scale, and exercises *control at a distance* with the same effectiveness, as presently achieved through extended international networks of wholly owned subsidiaries. This is an extreme example of where present trends may lead.

Chapter Three will analyse the internal changes inside multinational corporations as the result of the application of new technologies as well as the changes in the relationship between large European companies and their suppliers. Changes to be discussed will include the increased importance given to research and development, the reorganization of divisions as a result of converging basic technologies, the impact of new technologies on centralization and decentralization of production and decision making, and the question of how far the relationship between major companies and their suppliers shows a trend towards what might be called Japanization.

Japanization is also a key word for an analysis of changing labour relations. Labour relations will be dramatically affected by the application of new labour-saving technologies, especially the manifold application of microelectronics, and related organization concepts. On the one hand, much less labour will be needed in the future. This will not only be due to a far-reaching automation of production (and to a certain degree also of research and development, design and administration). It will also be due to a new design of products (less parts), the application of new materials (which will reduce attrition and cause a reduction of maintenance jobs) and the introduction of new management concepts. On the other hand, a smaller share of the remaining workers will stay employed in the multinational corporations themselves. Most of them will, instead, work with suppliers. These will not only not enjoy the privileges of the employees of the multinationals, they will also be deprived of much of their influence on company decision making, because the companies supplying the multinationals will themselves only have a limited say with regard to *what* they produce, *how* to produce it and at which *rhythm*. These impacts of the application of new technologies by multinational companies on employment and the distribution of influence will be dealt with in Chapter Four.

Much of the restructuring actually taking place has been provoked by increasing international competition from the United States and Japan. American companies saw their competitiveness in traditional industries even more threatened than European companies and therefore had an even stronger incentive to specialize in 'high technology'. It is especially in this area that European industry is challenged by competition from the United States and Japan. Worries about a decline of Europe's international competitiveness very much concentrate on this field. We therefore analyse the international position of European multinationals in core technologies in Chapter Five. Our thesis is that the more diversified nature of European companies (in comparison, first of all, with their American competitors) is one of the reasons for the fact that they have been less innovative in some areas. However, this very disadvantage could be turned into an advantage, if European multinationals were able to capitalize on the fact that basic technologies of different industrial branches increasingly converge. Given their higher degree of diversification, they should be better able to use technologies developed in one field quickly in other fields as well. They thus should be able to realize a *faster diffusion* of new technologies across industrial sectors, even where they have not been the first to introduce the original innovation itself. This, however, is a *management problem* that hardly seems to have been tackled by European multinationals, where the independence of divisions often stands in the way of an intracompany technology transfer.

International competitiveness does not only depend on the efforts of individual firms, but depends at least as much on the social context in which companies operate. Institutional and political factors play an important role. Governments have become important actors in technological development. In practically all industrial countries, governments have taken (additional) measures to speed up technological development and the diffusion of new technologies, since their macroeconomic objectives like economic growth, employment and an improved balance of payments depend to a large extent on the degree to which national economic actors master the new technologies. The strong emphasis put on technological development by all governments has led to a new relationship between large companies and the state (at the national level as well as at the 'supranational' level of the European Community). This new relationship between multinational corporations as important 'carriers' of core technologies and their respective governments is described in Chapter Six.

National governments have all started to support more or less the same technologies. They are caught in a kind of restructuring race, in which international competition among companies tends to turn into outright rivalry among states. The large corporations often play the role of 'national champions' on which national governments rely for the development of new technologies and the improvement of the country's export position. Especially in the case of European corporations in the microelectronics cluster, the companies' orientation on their national government as the major client and financier is still surprisingly strong—to such a degree that one may wonder how 'multinational' these European multinationals in fact are.

While international competition has stiffened and the increasing involvement of governments in technological development and the support of individual companies and industrial sectors leads to international political conflicts, we witness at the same time that fast technological development also leads to a *proliferation of international alliances* to speed up the acquisition of new technologies, products and overseas markets. These alliances are the main topic of Chapter Seven. Three questions are relevant for us in this context. First, we look at the development of *European technology programmes* and the resulting patterns of cooperation between European companies. Second, we compare these networks with the cooperation agreements that European companies have concluded with American and Japanese companies. It is often claimed that European companies tend to prefer cooperation with an American or Japanese partner to strengthen their position against European competitors, rather than cooperate with other European companies to improve their position with regard to competitors from overseas. We want to know how far this is the case and how far there is a real choice, in the first place, between cooperation with a European or a non-European partner, or whether these cooperation agreements serve different purposes. Third, we should like to know what the impact of this increasingly intense network of international cooperation agreements has on international competition. Will the main lines of competition be between European, American and Japanese companies? Or will there be more and more alliances comprising firms from all three continents, and will international competition first of all take place between different international alliances? The political implications of these possible alternative developments are obvious. Political conflicts among states can be expected to become much more intense in the first case than in the second. But more probable is that neither of the two alternatives will materialize in a clear-cut form, but that tensions in some field will go along with cooperation in others. This book intends to contribute to a first exploration of the resulting patterns of conflict and cooperation.

Chapter One

The Challenge of New Technologies

The world economy is actually undergoing a massive restructuring. This is a reaction to the economic crisis in the 1970s and the early 1980s. Western Europe has been more affected by the crisis than either the United States of America or Japan. While economic growth (of Gross Domestic Product) in Japan still averaged 4.4 per cent in the 1974–84 period, the United States only realized 2.7 per cent and Europe only 1.8 per cent. And while official unemployment never reached more than 2.7 per cent in Japan and 9.5 per cent in the United States, it climbed to 10.8 per cent in the European Community.[1]

Different economic measures have been suggested to counteract the crisis. While *Keynes* was the most cited economist of the *1960s*, *Friedman* probably dominated the discussion in the *1970s*, but his name became replaced by *Schumpeter* in the *1980s*. Schumpeter in the 1930s had elaborated *Kondratieff*'s[2] theory of long waves of economic development, which has become very popular again. According to this theory as interpreted by Schumpeter,[3] a new set of basic technologies (as the steam engine or the railway system) is necessary to start a new long-term upswing of the economy. After a while, these technologies will then become mature, most of the investment to exploit them will have been done, and the positive stimulus on the economy will whither away, leaving way to stagnation and even recession.

According to this stream of analysis, the postwar boom had been based on massive investment in the automobile industry and the petrochemical industry as well as the infrastructure necessary to make large-scale use of the products of these industries.[4] But this investment had to level off because a stage of maturity had been reached and return on investment consequently declined. Classical Keynesian measures would not help to turn the tide. This is not only because public deficits are already very high as a result of the crisis, and not only because international interdependence has become so intense that national efforts to stimulate the economy have

---

[1] OECD (1985a).
[2] Kondratieff (1926), Kleinknecht (1984).
[3] Schumpeter (1939).
[4] See Mensch (1979).

become self-defeating, as they tend to stimulate imports more than anything else. According to Schumpeterian approach, classical measures would not work because investment opportunities have become exhausted. Monetarist measures could not provide a solution either. While effective to cut down inflation, they alone would not get the economy going again. What is needed, according to a Schumpeterian vision, is the development of new basic technologies that can bring about a new long-term economic upswing.

To have this effect, the new basic technologies would have to lead to a host of new products and thus open up new investment and growth opportunities. Or they would have to make the production of existing products so much cheaper that totally new markets would become available to them. But this is not enough. The new basic technologies would also have to provide an answer to those problems that have beset the last phase of expansion based on the technologies that dominated the postwar boom. These problems will be listed below. Later on in this chapter we shall try to see how far the new basic technologies make a contribution to overcoming these problems.

1.1. OBSTACLES TO FURTHER ACCUMULATION

Six obstacles to further accumulation will be discussed here. These are: (a) increasing cost of labour, (b) increasing capital intensity, (c) increasing pollution, (d) increasing energy consumption, (e) increasing consumption of other raw materials and (f) increasing inflexibility of the production apparatus.

(a) *Cost of labour*   Throughout the postwar period, wages in the industrialized countries had grosso modo developed at the same rate as (or somewhat less than) productivity increases. This has given shape to patterns of bargaining and mutual expectations that led to continuing wage increases even when productivity growth slowed down. Wages started to increase faster than productivity in a period (in the late 1960s and early 1970s) when competition from a number of low-wage 'newly industrialized countries' first became felt. Overall market penetration by products from these countries never reached more than a marginal level,[5] but its fast increase in a small number of sectors (clothing, consumer electronics) underlined the potential for a massive shift of industrial capacity towards countries hitherto regarded as underdeveloped. Intense competition from

[5] OECD (1979).

low-wage countries (and lower trade barriers as the result of the Kennedy and the Tokyo rounds of the General Agreement on Tariffs and Trade (GATT) negotiations) made a high wage level look very dangerous for international competitiveness.

Since labour intensive production processes were the first to experience the competition from developing countries, a shift towards more capital intensive production was envisaged. Capital intensive production processes, however, soon became candidates for production transfer as well, because a low wage level was not the only attractive feature of newly industrializing countries. These countries were also comparatively free of regulations that restricted the intensive use of expensive machinery. While shiftwork decreased in the industrialized countries, developing countries offered the possibility to use expensive equipment around the clock, seven days a week, all year long. Only *knowhow* intensive production therefore remained rather protected from increasing competition from the Third World.

(b) *Increasing capital intensity* Rising wages stimulated further *automation*. Automation in the 1960s and 1970s, however, meant *fixed* automation, i.e. an investment in equipment that could only produce *one* specific product. In order to make this investment profitable, very large numbers of identical items had to be produced by the same machinery. The optimal factory size therefore increased. This, however, created a number of problems:

— In order to undercut production costs of competitors, companies were forced to build even larger factories in order to realize higher economies of scale. When the market did not expand as expected, or when several competitors realized massive investment plans at the same time, overproduction was the result. The amplitudes of business fluctuations thereby increased.
— Larger factories often had to be split up, because they put too much strain on regional labour markets and the environment or were not opportune for political or other reasons: companies might want to *produce* in several different markets (and not only serve them with exports from another country) either as a sign of goodwill or to remain in closer contact with developments there. Heavy currency fluctuations from 1973 onwards also obliged some companies to spread their assets in order to better match assets and liabilities in different currencies. A more decentralized company structure, however, brought additional costs (more intracompany transport, higher cost of communication and coordination, higher inventories of intermediate products).
— Capital had been rather cheap in the 1960s and 1970s. Real interests sometimes even became negative in the 1970s. When interest rates all of a sudden rose steeply, companies that had financed their expansion

from outside sources became burdened with high interest rates, often in a situation when their cashflow stagnated or even declined.

(c) *Pollution*   Fast economic growth often led to increasing pollution of the environment, eventually approaching a critical level when irreparable damage was threatened if no prompt actions were undertaken. People fleeing polluted areas and settling outside but commuting back and forth every day only aggravated the problems. This prompted government regulation which made additional capital outlays necessary to meet the new standards.

(d) *Energy consumption*   The economic growth realized after the Second World War has been extremely energy intensive. Between 1950 and 1978, energy consumption increased 2.6 times.[6] The convenience of oil as the major source of energy had made many industrialized countries dependent on large-scale oil imports from a few countries. When these countries seized their increased bargaining power and used it to enforce higher oil prices, the vulnerability of the chosen form of industrialization became obvious and triggered measures to economize on energy.

(e) *Raw material shortages*   During the postwar economic boom, mankind (or that small fraction of mankind that lives in the highly industrialized countries of the Organization for Economic Cooperation and Development (OECD) area) had consumed as much of the world's non-renewable resources as earlier generations had used during the whole process of civilization in the centuries before. As in the case of energy, this has led to the necessity for huge investments in the exploitation of less rich or more difficult to access deposits and to a high dependence on imports. With increasing discrepancies between living conditions and aspirations in many countries outside the highly industrialized world and increasing unrest in many countries, access to necessary raw materials did not seem assured any longer. Price rises for some commodities in the aftermath of the oil price increase, too, pressed governments and enterprises in OECD countries to economize on raw materials.

(f) *Lack of flexibility*   Highly developed systems of labour relations, huge investments in large plants with fixed automation equipment, an intricate intracompany division of labour and unavoidably detailed environmental regulations have led to a situation in which the production apparatus has become rather inflexible. Changes would have to be incremental in order not to devaluate earlier investment. Besides, the very success of mass production at the same time has undermined the future viability of the productive structure that it has shaped. 'To the extent that consumers demand a particular good in order to distinguish themselves

[6] United Nations (1980).

from those who do not have it, the good becomes less appealing as more of it is sold. Consumers will be increasingly willing to pay a premium for a variant of the good whose possession sets them off from the mass.' Consumers are not only increasingly willing to pay a premium for non-standard products, they are also increasingly able to afford it: 'As mass production cheapens the cost of manufactured goods and raises real wages, consumers have the discretionary income to experiment with their tastes.'[7] Production of high-volume, standardized products therefore may run into problems.

Once the economic crisis deepened, it had its own effect on the differentiation of demand. While people directly affected by the crisis had to adapt their pattern of consumption, those who remained employed or found alternative employment in new expanding activities often experienced continued wage increases. This explains the often increasing demand for differentiated luxury goods at a time of crisis.

When profits started to rise again, investment in new capital goods increased faster than consumer demand. Professional equipment, however, often has a shorter life-cycle than consumer products. With a shift towards professional equipment (e.g. the much faster increase in demand for personal computers compared to consumer electronics) the average life-cycle shortens, and companies have to develop more flexibility to adapt to changing demand.

The problems described above pressed for change. New technologies were looked for that would reduce the cost of labour, limit the capital needs, have less hazardous effects on the environment, consume less energy and other raw materials and provide for more flexibility. 'Innovation' consequently became a keyword in companies and governments. New products should help to explore markets not yet challenged by competitors, and new production processes should help to reduce production costs and negative (and increasingly costly) side-effects.

1.2. THE EMPHASIS ON 'HIGH TECHNOLOGY'

Consequently, most companies, supported by national governments, try to leave traditional activities and try to diversify into advanced technology fields. This shift of activities by European multinational companies will

---

[7] Sable (1982), p. 199. On positional goods, which are valued because few possess them, see also Hirsch (1978), pp. 27–54.

be described in the next chapter. The 'high technology' sector, however, is an extremely diffuse field. If one takes R&D expenditures as a measure, the industries comprising the high technology sector of the economy would be[8]:

Aircraft and parts
Computers and office equipment
Electrical equipment and components
Optical and medical instruments
Drugs and medicines
Plastics and synthetic materials
Engines and turbines
Agricultural chemicals
Professional and scientific instruments
Industrial chemicals

These industries play a key role in the economy. In the United States, for example, they had a rate of growth of real output which was more than twice that of total US industrial real output. During the 1970–80 period labour productivity of these industries grew *six* times faster than that of total US business.[9]

It would not make sense, however, nor would it be practical to deal with all these sectors in this study. In this study, we do not want to deal with industries, but with specific *technologies*. And in order to be selected for incorporation into this study, these technologies would have to form the potential basis for a new long-term economic upswing. We, therefore, are interested in technologies that

—will lead to many new products;
—will have a strong impact on production processes;
—are applicable in *many* sectors of the economy;
—would not intensify but instead ease the problems brought about by the postwar economic boom (see above).

Technologies that satisfy these four criteria we shall call *core technologies*. Two clusters of core technologies are the *microelectronics cluster* and the *biotechnology cluster*. Around these basic new technologies, new 'technology webs'[10] are being created. These basic technologies find

---

[8] See US Department of Commerce (1983a), p. 3 and Appendix A.

[9] US Department of Commerce (1983a), p. 3.

[10] See Roobeek (1987b). There is a broad literature on technological innovation and the role played by *basic*, or *general* or *generic*, innovations; see for instance Freeman (1982). An interesting study very much related to the concept of clusters chosen by us is of CPE (1985), *Grappes technologiques*.

application in many industries, and these applications will be described below in more detail.

It is due to this concept of 'core technologies' that we will not deal with most of the industries that are described as 'high tech' industries. First, we shall not deal with 'industries' as such, because firms in rather traditional sectors can spend a lot of effort to get a strong foothold in core technologies, while firms in highly research-intensive industries can concentrate on technologies which are hardly of any importance outside the industry in question. We are not going to deal with medical technology nor with traditional optical technology. We do not analyse progress in pharmaceuticals in general, nor in professional and scientific instruments. And we even leave out most of the space and aircraft industry as well as the nuclear power plant sector. The reason is that although they are important appliers of 'core technology', they themselves are not in the forefront of the development of these technologies. Another reason is that most of the companies in the above given fields (e.g. optical industry, professional and scientific instruments, space and aircraft industry) are not really multinational companies.

In the rest of this chapter, we first describe the two clusters of core technologies and how they have fundamentally changed the relations between different industrial sectors. We then discuss the contribution to a solution of the above mentioned problems of accumulation that core technologies may make.

1.2.1. Microelectronics: the present revolution

In postwar economic development the performance of the electronics industry, compared to all other sectors, has been extraordinary. It has been the only sector combining a growth in the productivity of *both* labour and capital.[11] This has made (micro)electronics the most promising technology of the last three decades.

Many innovations answered military needs, especially in the United States which has been the breeding ground for most innovations in microelectronics.[12] The importance of a fairly stable public procurement market cannot be underestimated since most microelectronics technologies initially were not very attractive for immediate commercial applications and had to be produced at great cost. The first prototypes of transistors, numerical controlled machine tools, and the first computers were jointly developed with the manufacturers and bought by the Pentagon.

---

[11] Kaplinsky (1985b).
[12] See Halfmann (1984).

The development process of microelectronics is further characterized by a continued miniaturization of devices and an increased number of components put on a single device which eventually led to a rapid decline in the number of building blocks for each end-product. This could only happen due to the process of *digitalization* in microelectronics.[13]

Semiconductors, robots, computers, telecommunications equipment, software together with Computer Aided Design (CAD) equipment are the most important parts of the microelectronics or information technology cluster. The first and most central core product is the *semiconductor or 'chip'*. Semiconductors have been called the 'crude oil of the 1980s'. Since the early 1970s the number of components which could be integrated on a single chip has doubled almost every two years. From small-scale integration with less than 100 components per semiconductor, via medium-scale integration and large-scale integration, we are now passing the stage of the sixth generation chip, the Very Large Scale Integration (VLSI) chip with more than 100 000 components on it. The pace of developments in the semiconductor industry, although extremely sensitive to business cycles, is likely to affect progress in all other high technology industries in the future.[14] Semiconductors form direct and strategic inputs for all other microelectronics products ranging from consumer electronics to robots and telecommunications equipment. In most areas of information technology and capital equipment applications, semiconductors are used to raise the reliability of the products they serve and to reduce capital requirements (lower energy consumption, less material used). With only a small increase in integrated circuits used in 20 inch colour television sets, for instance, a major decrease in separate components could be achieved, also leading to a 50 per cent reduction of energy consumption per television set.[15]

The development of the microprocessor in 1971 by Intel corporation was an important innovation since it resulted in a device which could perform different logical functions which other semiconductors cannot do: they can only be used for the function they are designed for. Microprocessors can be reprogrammed and form the central 'brain' of a computer. In fact, the present generation of 32-bit 'supermicroprocessors' have the capacity of previous generations of (large) mainframe computers being able to execute 1 million or more instructions

---

[13] All ingoing information is transformed into binary codes (either a 'one' or a 'zero', a logic gate being switched 'on' or 'off') which consequently can be processed at high speeds. The capacity of a chip is described in its number of 'bytes', a byte being a grouping of 8 bits of information taken together.

[14] See *Japan Economic Journal*, 29 June 1982.

[15] See Philips, Microelectronics: the pivot of an industrial revolution? July 1979; cited in Franko (1983), p. 94.

per second on a single chip. Within the category of semiconductors, microprocessors—next to memory chips—therefore are the most strategic components. In factory and office automation, the microprocessor can act as the central processing unit of individual robots, of the electronic drawing boards of the designers, even of the whole computer system, providing it with the framework for the desired flexibility at various levels.

While chips are the crude oil of the information society, the group of *telecommunications equipment* products is supposed to serve as its central nervous system. Although the telecommunications equipment industry has existed for around a century, it is far from being a 'mature' industry. The increased capability in telecommunications to transform analogue into digital signals and vice versa in the course of the 1970s caused a major transformation of the whole industry itself. It meant that computer technology could be—and had to be—incorporated into switching equipment. Further progress in transmission technology (coaxial and fibre optical cables, satellites) escorted these developments. Once broadband capacity is diffused, as a result, digital telecommunications networks can connect not only the traditional telephone equipment, but also computer terminals, television sets, machine control panels, telecopy machines, etc. For the large volume of binary information transmitting not only voice but also moving pictures and the speed and reliability of information transport between these different terminals, it becomes vital to increase the capability of the transmission equipment from 'narrowband' to 'broadband' and substitute electric impulses by optoelectronic signals. The fast development of new telecommunication technology and the creation of their own internal international communication networks by multinational corporations put the traditional monopoly position of national postal and telecommunication authorities (PTTs) and their prime suppliers under great pressure and can lead to far-reaching institutional changes (see Chapter Six).

A third category of core products are the consecutive generations of *industrial robots*, which can be typified as the centrepiece of flexible automation. This will certainly be the case once the third[16] generation of more 'intelligent' industrial robots (combining powerful processing units with vision system and tactile sensitivity) will have found their way into industry. Most models are still on the drawing boards. With the advancement of robot technology, the capacity of robots to perform ever more complex tasks depends increasingly on the development of the

---

[16] Manipulators and sequential robots are often regarded as the *first* generation of robots, which have limited flexibility as compared to the *second* generation of robots (playback and numerically controlled robots). Cf. OECD (1983b), p. 16.

electronic command structure, whereas the mechanical parts account for less and less of the total cost of robots produced.

In the next core area, *computers*, a fourth generation is coming up, although most of the discussion already centres on the fifth generation of computers.

The most important development in the 1970s was the substitution of large parts of the central processing units in the old computer architecture by only one device, the microprocessor. As a result smaller and cheaper versions could be produced with the same processing power as first and second generation mainframe computers. This very quickly led to the birth of 'personal computers' and home computers which enhanced the pace of diffusion of computers into households and offices and on the factory floor.

The first four generations—from electronic vacuum tube, transistorized computers, integrated circuit computers to very large-scale integrated (VLSI) computers—are all based on the basic architecture of John von Neumann, which contains a central processor, a memory, an arithmetic unit and input–output devices. They are operated in a serial (step by step) manner.[17] The fifth generation is supposed to operate with parallel information circuits and handle not only numbers but is also capable of performing symbolic inference, i.e. simulating a process of reasoning. The change from fourth to fifth generation computers will be revolutionary. Radically new chips, new software, and advances in artificial intelligence will be needed to make this leap possible.

A fifth area of special interest is *Computer Aided Design (CAD)*, basically the integration of minicomputers, digitizing drawing boards and visual display units.[18] The development of interactive computer graphics has greatly profited from progress in computer, semiconductor, tube and software technology. Interactive graphics were first used in the 1950s in radar systems operating on a mainframe computer. Since then progress has been made in the processing of ever more advanced and complex pictures, i.e. 3D images.

Present generations of CAD turnkey systems no longer have to be hooked up to a large computer to perform well. Their own computing power suffices for an increasing number of tasks. This makes it possible for many smaller engineering companies to install CAD equipment.[19] CAD clearly saves drawing labour by automating all routine tasks. In addition, computerized testing of models in the design stage and advance cost calculations become feasible. CAD is not only applied to the

---

[17] See Feigenbaum and McCorduck (1984), p. 15.
[18] Kaplinsky (1984), p. 44.
[19] See Arnold (1984), p. 4.

development of new products, but also to the design of whole plants. CAD makes a digitalization of the design input possible, which is a precondition for a further integration of development, production and administration in any firm. Without CAD, next generations of large-scale integrated semiconductors and computers are unthinkable. CAD, therefore, increasingly also becomes a critical input for progress in most other core parts, not only of microelectronics, but also of biotechnology (for instance, in protein engineering).

Finally, a special category of core technologies is *software*. Constraints in the application of new technologies are largely due to operating systems, which are the control programmes for the central processing unit or microprocessor. The operating systems depend therefore on the power of the processor. The most important operating systems are the IBM mainframe systems like VHS and CICS, and the Digital Equipment systems like VAX and RSTS. Widely used micro operating systems are, for instance, Apple's DOS (Disk Operating System) or Microsoft's MS–DOS in the area of personal computers.[20] A new powerful operating system developed by American Telephone and Telegraph (AT&T) in the first half of the 1980s is the UNIX system which seems to get industry-wide acceptance.

Next to the 'systems software', which is very much linked to the progress made in hardware developments, a whole range of applications software has evolved.[21] Advancements in software language have resulted in more 'user friendly' versions which enable the operator to work in a conversational mode with the computer. In general, the software content in all of the previously mentioned core technologies has grown rapidly. The CAD industry, the telecommunications industry or the computer business to an increasing extent can be analysed as software, or 'thoughtware', industries. Software has become one of the major bottlenecks in keeping track of developments in larger-scale microprocessors and semiconductors. It has a crucial function in the further integration of core technologies.

These six product areas form the core technologies of the microelectronics cluster. The technologies as such often cannot be viewed as revolutionary. However, the *combination* of parallel developed technologies — with important cross-fertilization — results in a revolution of sorts in which the total is more than its parts. The group of technologies that the microelectronics cluster contains, affects every thinkable user industry from printing to textiles production. Some even claim that around

---

[20] Davidson (1984), p. 120.
[21] The International Software Directory classifies 107 groups of applications. (1985d), p. 24.

80 per cent of the Western world's production in fact has become *dependent* in some way on the electronics sector.[22]

1.2.2. The next revolution around the corner: biotechnology

At the centre of the second cluster, we find *biotechnology*, 'the application of scientific and engineering principles to the processing of materials by biological agents to provide goods and services'.[23] This definition excludes the age-old use of the productive capacity of microorganisms for baking bread, making cheese, brewing beer and producing wine (the so-called *first generation* of biotechnology). It would include, however, more recent applications of fermentation technology as in the production of penicillin. The development of penicillin during the Second World War has led to a renewed interest in fermentation processes, to the production of new antibiotics and to a wider use of enzymes (proteins produced by living cells that act as catalysts that mediate and promote chemical reactions without themselves being destroyed in these processes). These developments constitute the *second generation* of biotechnology. Their significance has been threefold:[24] (a) They stimulated intense interest in microbiology, i.e. the properties of enzymes or of microorganisms such as fungi and bacteria. (b) This in turn led to the use of mutations and selection of strains to achieve very substantial improvements in yields and production efficiency of the microorganisms. (c) This was accompanied by considerable refinements in fermentation technology, the most notable of which was the discovery of methods of immobilizing enzymes. This meant that enzyme catalysts could be used over and over again instead of being lost with each batch processed. This has made the shift towards *continuous* fermentation processes possible.[25]

The great breakthrough of biotechnology, however, was only reached in 1973, when researchers at Stanford University succeeded in isolating part of the genetic code (DNA) of a cell to integrate it into the DNA substance of another organism, thus transplanting a specific characteristic into this organism. This has opened up vast opportunities for *genetic engineering*. Technologies that alter the genetic material are often called *third generation* biotechnology. In order to be used

[22] According to Philips's president Dekker, in *International Management*, October 1985, p. 32.
[23] Bull *et al.* (1982), p. 21.
[24] See Sharp (1985b).
[25] Sharp (1985b), p. 15.

successfully, however, they have to be combined with 'second generation' technologies.

Biotechnology, therefore, is not just one technology, it is a *set of technologies*. At least four different technologies have been at the basis of the extremely rapid progress that has been achieved during the last fifteen years.

The first and most debated is *recombinant DNA technology*. This is the technology to separate genes responsible for specific characteristics of an organism, and to introduce them into the DNA code of another organism. In this way the ability to produce a valuable substance, for example, can be introduced into a microorganism that multiplies quickly, thus enhancing the productive capacity of microorganisms in a spectacular way.

The second is *cell fusion*, the artificial joining of cells, by which the desirable characteristics of different types of cells can be combined into one cell. An important use of this technology has been made for the production of so-called *monoclonal antibodies* which can be used for many purposes, including the diagnosis and treatment of disease, the purification of proteins and the production of biological sensors.

The third and somewhat less glamorous technology is *enzyme technology*. Without enzyme technology, recombinant DNA would not have been possible, because enzymes are needed to isolate genes from the DNA of any species. In most industrial biotechnology processes, the use of enzymes is crucial to bring about the reactions desired and to recover the substances wanted out of the fermentation slurry.

The fourth is *bioprocess technology*, a set of technologies in itself which allows the scaling up of production of useful substances. This englobes modern fermentation technology, the immobilization of enzymes, and a host of downstream technologies to recover the valuable substances from large quantities of cellular components, nutrients, wastes and water which result from the fermentation process. It is very often this final stage which is decisive for whether a given process which is *technologically* feasible also proves to be *commercially* viable.

Both microelectronics and biotechnology form clusters of overlapping core technologies.[26]

The six elements on the microelectronics side and the four on the biotechnology end in the last 10 to 40 years have known parallel developments in which technological inventions in one area stimulated or

---

[26] A good illustration of the strategic nature of both technological clusters is the fact that 'of the sixteen American firms that built manufacturing facilities in Japan during the first half of 1982, ten were in the business of making advanced semiconductors, and four in biotechnology and fine chemicals', *Japan Economic Journal*, 21 September 1982, p. 3; quoted in Reich (1983), pp. 265–6.

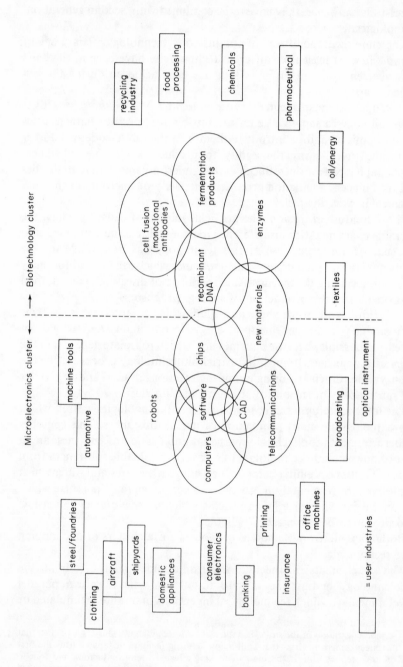

Microelectronics cluster ←——→ Biotechnology cluster

steel/foundries

clothing

aircraft

shipyards

domestic
appliances

consumer
electronics

banking

printing

insurance

office
machines

machine tools

automotive

robots

computers

software

CAD

chips

telecommunications

broadcasting

optical instrument

new materials

recombinant
DNA

cell fusion
(monoclonal
antibodies)

fermentation
products

enzymes

textiles

oil/energy

recycling
industry

food
processing

chemicals

pharmaceutical

☐ = user industries

Figure 1.1 Main technology clusters in product innovation

made possible progress in other areas, first within the own cluster, but increasingly also across the clusters. Figure 1.1 gives a graphical illustration of the main technologies comprising the two basic clusters and the wide range of user industries directly affected by progress in the core technologies.

New materials, as a third category of product innovation, have been added. They include: specialty steels, composites and polymers, plastics and ceramics. The innovations coming from progress in new materials have formed important inputs for innovations in microelectronics as well as biotechnology. Chip technology, for instance, has been identified with silicon, the material of which the first generations of chips are made.[27] But the potential of silicon proved too limited for further miniaturization. Advancement in chip technology at the moment is linked with advancement in the use of other materials such as gallium arsenide (GaAs) or optical material combined with lasers, which make faster and smaller circuits on a submicron level possible. The use of optical fibres in telecommunications transmission equipment is another clear example of new materials as an input to further developments in the microelectronics cluster. The automation of the production process, too, can often only progress if new materials are used that can be more easily manipulated and transformed by machines.

New materials are also vital for the further development of biotechnology. New materials for membranes, for instance, could revolutionize downstream processes in fermentation and be decisive for the profitability of the biotechnology route of production. On the other hand, new materials are partly the consequence of developments in biotechnology and can be the cause of advancements in microelectronics. New materials, therefore, in some areas can form the *link* between the biotechnology and the microelectronics clusters, although the group of technologies certainly has also a dynamism of its own.

1.2.3. Increased convergence between and within the clusters

It has been noted that there is a considerable interaction between product developments in the different core technologies within the respective clusters. Although there has not been any major influence of developments in the biotechnology cluster on microelectronics so far, the use of microelectronics technology in biotechnology research and in fermentation processes has been substantial.

---

[27] See, for instance, the reference to 'Silicon Valley' in California or 'Silicon Glen' in Scotland.

It is in the field of bioprocess technology where microelectronics and biotechnology first meet. Fermentation technology has acquired a new quality because the application of microelectronics makes it possible to control and steer the fermentation process much more accurately than in the past. Another field where biotechnology and microelectronics overlap is that of the research process itself. Modern biotechnology research would hardly be possible without the use of microelectronics. Specific CAD equipment has been designed that helps biotechnologists to map the structure of protein molecules and to better understand the characteristics of these molecules. Microelectronics-based *'gene machines'* help to use this knowledge. They help to find specific gene sequences and facilitate the manipulation of genetic information.[28] The field of *bioinformatics*[29] furthermore includes databanks that contain information on micro-organisms as well as on literature and research results in the field. The 'biochip' eventually will make use of the ability of cells to store a vast amount of information in the DNA code.[30] It may not only be used as a memory chip. The capacity of cells to process information may eventually be applied as well in computers that will provide 'artificial intelligence'. It sounds very futuristic, but research is actually done to lay the basis for an integration of biotechnology and microelectronics.[31]

It is predominantly in the microelectronics cluster that technologies are converging and become more and more linked.

The growing integration of telecommunications and computers, of robots and microprocessor technology, of CAD and robots, has a number of consequences for the firms investing in these product innovations:

(a) *Different technologies have to be mastered at the same time;* a firm traditionally operating in, for instance, telecommunications has to invest also in computer, CAD, chips and software technology since it is the only way of keeping pace with the digitalization of its industry. The direct effect is a blurring of traditional sector boundaries.

(b) This results in *higher R&D expenditures* for developing the latest generation of core products.

---

[28] See OTA (1984a), p. 87.

[29] See the special programme on *bioinformatics* of the BMFT (the Federal Ministry for Research and Technology), Bonn, 1985.

[30] 'DNA can be thought as a library that contains the complete plan for an organism. If the plan were for a human, the library would contain 3,000 volumes of 1,000 pages each. Each page would represent one gene, or a unit of heredity, and be specified by 1,000 letters', OTA (1984a), p. 33. Such a tremendous library is in *every* human cell.

[31] The Department of Defense in the United States started funding biochip research beginning in the fiscal year 1984 at $3 million to $4 million for 5 years. A few large electronics companies in the United States (Westinghouse, General Electric, AT&T and IBM) have small in-house programmes in this area. Japan, France, the United Kingdom, and the USSR have indicated interest in biomolecular computers. OTA (1984a), p. 254.

(c) For these massive R&D investments *growing worldwide sales* are needed to give the company enough 'critical mass'.

(d) Firms easily *move into areas of the most profitable applications* which they might produce with a competitive advantage in time and production knowhow setting the standard for the end-product's specifications.

In the move towards more integration of core technologies, product and process innovations become closely linked. Process innovations will almost revolutionize all fields of production activity. Parallel to the division of the economy into three productive sectors, we can distinguish three clusters of process innovation which together form

(a) The Factory-of-the-future[32]
(b) The Office-of-the-future
(c) The Farm-of-the-future

This is illustrated in Figure 1.2. Other areas of importance might be (d) the Household-of-the-future, and (e) the Battlefield-of-the-future in which 'Star Wars', robotized warfare (with so-called 'drones') and chemical warfare are practised. This book, however, concentrates on the

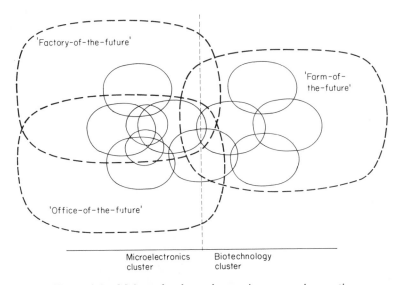

Microelectronics   Biotechnology
cluster   cluster

Figure 1.2   Main technology clusters in process innovation

[32] The General Electric Co. in the USA rather refers to the 'Factory-with-a-future'. It has to be kept in mind that speculations on the concept of a fully automated factory were already being made in the 1950s. This was particularly stimulated by the US Air Force in cooperation with the same General Electric Co. Cf. Noble (1984).

1 Increasing level of integration between the Office and the Factory. Relatively low level of integration between microelectronics and biotechnology clusters (1990)

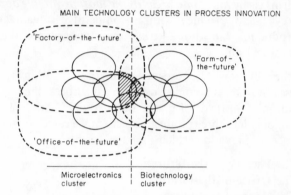

MAIN TECHNOLOGY CLUSTERS IN PROCESS INNOVATION

'Factory-of-the-future'

'Farm-of-the-future'

'Office-of-the-future'

Microelectronics cluster

Biotechnology cluster

2 Continued integration of Office and Factory, accompanied by integration of Factory and Farm (2000)

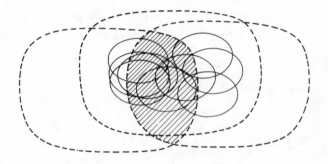

3 Continued integration of all three spheres (2030?)

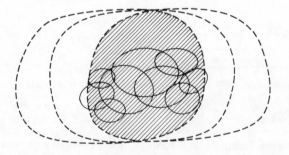

Figure 1.3   Growing integration of the technology clusters

development of productive activities and will only occasionally pay attention to new consumer goods and to the development of the means of destruction.

At the moment the overlapping area (the core of these core technologies applied in all these sectors) is still small. If we extrapolate the present tendencies towards increased technological convergence, integration is bound to proceed into a situation in which especially office and factory automation become closely intertwined and based on the same technologies. The digitalization of the farm will proceed also. On the base of growing use of biotechnological materials and techniques in the microelectronics cluster, the sphere of the Farm-of-the-future might integrate with that of microelectronics. The final outlook may be a large technological core region in which biochips, biocomputers and the like are used for very advanced and very complex applications in offices, factories and farms at the same time. Firms which have been able to move into this core area will control the strategic inputs for their own production process and for most other applications in industrial commodities, in services as well as in food and beverages production. Figure 1.3 gives a graphical impression of this process.

1.3. COPING WITH ACCUMULATION PROBLEMS

It is not enough that the core technologies lead to a large number of new products. It is as important that they make new production processes possible which are cheaper on the one hand, and which, on the other, promise to overcome the bottlenecks that have throttled postwar accumulation.[33]

The technologies described make production cheaper, because they help to economize on labour[34] and on capital. They help to save raw materials and to reduce energy consumption. They are also regarded as having less negative effects on the environment (and can be used to control, limit and clear away pollution caused by traditional industrial activities). Core technologies are also expected to provide more flexibility in all phases of the production process. The technologies described thus fulfil the conditions formulated earlier in this chapter. It is because of these characteristics that the technologies that have just been presented can be called 'core technologies'. How these technologies are expected to

[33] See Roobeek (1987b).
[34] The impact of the application of these technologies by European multinationals on employment will be dealt with in detail in Chapter Four.

Table 1.1   *Reduction of the amount of labour necessary to produce successive generations of television sets. (K6–K12 are different types of Philips TV sets)*

From K6 to K7 and K8	− 20%
From K8 to K9	− 40%
From K9 to K11	− 10%
From K11 to K12	− 25%

Source: Rodenburg (1980), p. 366.

contribute to a solution of the various bottlenecks identified above is described below.

1.3.1.   Core technologies save labour

The application of microelectronics clearly reduces the amount of labour spent in production. On the one hand, microelectronics is used in the *products* themselves where a single chip is often a substitute for a large number of moving parts which needed a lot of time to be manufactured and assembled. A good example is the fast declining amount of labour needed to produce successive generations of television sets between 1967 and 1978 (see Table 1.1). For many other electrical and electronic products (like cash registers, electrical typewriters and even washing-machines), a similar development has taken place.

On the other hand, microelectronics is used in the *means of production*. Examples are word processors, numerical controlled machines and robots. It is estimated that the robots introduced hitherto have replaced two or three workers, and that robots of the second generation will replace four to six workers. In some cases, up to ten jobs will be lost.[35] Concrete examples of European multinationals show somewhat less dramatic results. The RENAULT plant at Douai, for example, used 125 robots in the early 1980s which were estimated as being 20 per cent more productive than manual labour.[36] The 'employment' of a robot costs about $6 per hour in the United States, whereas the wage of a worker in the automobile industry doing the same job would be about $19 per hour.[37] The labour cost that can be cut by the use of robots is not only the cost of labour in the manufacturing process. Robots are steered by digital information and can thus easily be linked up with storage facilities for raw material

---

[35] See Chapter Four.
[36] *Die Welt*, 26 October 1983.
[37] OECD (1983b), p. 75.

and half-finished products as well as with 'downward' administrative activities such as billing (see Chapter Four for 'intersectoral automation'). In this way, economies can also be made in employees' work in the administrative sector.

Better communication and coordination by *telecommunication* facilities help to boost productivity in a number of ways:

(a) Better communication helps to direct different workers to the place where their activity is needed most.

(b) Better logistics[38] help to reduce time that is spent on waiting for material to arrive.

(c) Telecommunications helps to reduce transport activities by contributing to a decentralization of production processes.

(d) The transmittance of information can often be a substitute for the transport of goods and people. By bringing people in contact with each other over large distances, it becomes less necessary for them to displace themselves and thus reduces the time spent on travelling and commuting.

(e) By making all relevant information available at the place where it is needed, work is less often interrupted until a specific question is answered or a given problem is solved.

(f) By increasing the control of production activity, telecommunication helps to avoid things going wrong and having to be done all over again.

1.3.2. Core technologies save capital

The same technologies not only help to economize on labour, they also help to cut down capital requirements. This happens in a number of ways:

(a) The introduction of microelectronics has brought a dramatic change to *price/performance ratios* for capital goods. They are best illustrated by the development in the field of computers themselves, where personal computers have become as powerful as mainframe computer used to be a few years ago — at a fraction of their price.[39]

(b) Beside changes in the price/performance ratio of capital goods, capital goods have become *reprogrammable*. Thus, the same machine can perform several functions. It can therefore often *replace a number of other machines* and thus reduce the total amount of machinery used.

---

[38] *Blick durch die Wirtschaft* (Frankfurt), 31 October 1983.
[39] If automobiles had shown a price decline as similar to that of memory capacity in computers, during the last 20 years, a Rolls-Royce would not cost more than 50 cents now.

(c) Reprogrammable machinery can adapt to new product lines. This *stretches depreciation* out over a longer period. In the motor industry, for example, 60 per cent of the investment can be saved with a robotics manufacturing system when changing models, whereas fixed automation has to be more or less written off.[40]

(d) Automatic machinery can be used *around the clock*. It thus allows a much more intensive use of machinery (which reduces the machinery stock necessary).

(e) The use of automatic machinery (and multipurpose machinery that reduces the total number of machines in use) *reduces the factory space necessary* because machines can be 'packed' more densely if less room is needed in between for workers, and if new systems of logistics reduce the amount of raw materials and intermediate products that have to be stockpiled on the factory floor (actually often about one third of all factory space).[41]

(f) Better logistics and 'communication' between the machines can optimize the path any piece of work has to follow from one machine to another, thus *optimizing machine use* and minimizing the time machines stand idle.

(g) Microelectronics further *reduces downtime* by having machine built in a modular way that allows the quick exchange of parts if repair is necessary. Microelectronics also has led to the development of *self-diagnosing* and even *self-repairing* systems, which contribute to a further reduction of downtime.

(h) *Biotechnology* will make it possible to substitute a broad range of petrochemical processes. Whereas these take place at high temperature and under high pressure, biotechnological processes take place at low pressure and 'normal' temperature. Since capital equipment will not have to stand such extreme conditions, it can be much cheaper.

(i) Capital cost can be reduced even further by the use of *new materials*. They will reduce wear and abrasion and thus will contribute to keeping equipment going for longer periods.

1.3.3. Core technologies save raw materials

Core technologies help to economize on capital equipment, but also on other forms of capital outlays. They help to reduce working capital by reducing the amount of raw material needed in production and by streamlining the flow of materials to minimize stocks.

Core technologies help to realize these gains in the following ways:

---

[40] *Financial Times*, 16 April 1982.
[41] Bylinski (1983).

(a) *Miniaturization* and the reduction of the number of parts made possible by microelectronics help to reduce the material involved in many products.

(b) The use of material can be reduced where computers are used to calculate the *optimal cutting* of raw materials (e.g. textiles, leather) in such a way that waste is avoided as much as possible.

(c) Since flexible automation will make it possible to adapt production lines quickly to new products, it will no longer be necessary to have large stocks of different parts because they can be easily produced as they are needed. This helps to reduce inventories.

(d) Telecommunication links to dealer networks can keep central management informed of actual market developments and help to *adapt production in time to fluctuations of demand*. This will reduce the amount of working capital immobilized in unsold stocks.'

(e) Communication links to suppliers help to introduce 'just-in-time' delivery which can slim down inventories considerably.

(f) New technologies in general increase the substitutability of different raw materials. It therefore makes it possible at any moment to select the cheapest raw material from an increasingly broad spectrum of alternative raw materials.

(g) New materials also comprise new raw materials which are much cheaper than the materials previously used. Optical fibres are cheaper than coaxial wires, and new forms of cement can be cheaper than the aluminium which is substituted by it.

(h) It is said that biotechnology will even lead to some 'dematerialization of production'[42] which implies that in general smaller amounts of raw materials will be necessary to realize the same 'use value' as in the past.

1.3.4. Core technologies reduce energy consumption

Specific raw materials are fossil fuels. It has been mentioned that postwar accumulation was accompanied by a more than proportional rise in energy consumption. Rising energy prices have — rightly or wrongly — often been identified as the most important cause of the economic crisis of the 1970s. New technologies have helped to reverse the trend of ever increasing energy bills.

(a) Better means of telecommunication reduce the demand for energy as the *transmission of information is a substitute for the transportation of goods and persons*.

(b) Automation of production reduces the amount of energy needed to adapt the shop floor to the needs of human labour (light,

---

[42] See Sargeant (1984).

air-conditioning). It also leads to a more *continuous use of energy* by spreading energy consumption around the clock)

(c) Microelectronics devices themselves need less energy than mechanical devices. Since the number of moving parts is reduced, *friction is minimized* and additional cooling becomes less necessary.

(e) Since biotechnology processes take place at *normal temperature* and under normal pressure, less energy is needed for heating and compressing material to be processed.

(f) New materials often have been invented with an eye on economizing in energy consumption. They can be used for a *more efficient use of fuels* (e.g. ceramics in motor blocks which allow higher temperatures), for *better isolation*, and to be a substitute for traditional processes which were highly energy intensive (e.g. the separation of different materials from a fermentation brew through membranes where the liquid otherwise would have to be heated to evaporate).

1.3.5. Core technologies help to fight pollution

The new technologies are often associated with increased 'white-collar' employment, the white collars remaining white because the new technologies will reduce pollution. 'High tech' industries are regarded as the opposite of the 'smokestack' industries. This is by and large true, although semiconductor production itself has led to severe environmental problems in the areas where it is concentrated most, such as Silicon Valley. The *application* of microelectronics and biotechnology can reduce environmental problems in a number of ways:

(a) Better design and better optimized cuttings lead to a considerable *reduction of waste*.

(b) Microelectronics can help to monitor production sites and signal pollution at an early stage.

(c) Better telecommunications facilities make *decentralization* possible. By substituting the transport of people and goods by the transmittance of information, telecommunications contributes to *restrict traffic* and the pollution connected to it.

(d) Biotechnological production processes often do not use the same toxic and aggressive substances which are used in many traditional processes. They therefore cause less environmental damage (although new environmental hazards might be created by biotechnology).

(e) Where pollution has taken place, biotechnological methods can make an important contribution to clearing up the sites in question.

Table 1.2 Core technologies are expected to remove bottlenecks which restrict growth

Bottlenecks	Semiconductors	Telecom.	Robots	Biotech	New materials
Products	New consumer products; new capital equipment	New infrastructure; many new services	Products of higher quality	New pharmaceuticals; new materials; new plant varieties	New materials: polymers, plastics, composites, steels
Labour costs	Reduces number of parts to be assembled; allows flexible automation	Facilitates 'inter-sectoral' automation	Direct replacement of human labour	Less labour intensive production; indus-trialization of farm	Redundancy of certain professions (e.g. welding); increased possibilities of automation
Capital costs	Makes machines reprogrammable	Helps to optimize machine use	Reduction of factory space; use of equipment around the clock	Less costly equipment (not resistant to heat and high pressure)	More intensive use of equipment
Raw materials consumption	Miniaturization; optimal cutting	Better coordination reduces stocks	More precise work causes less scrap	Increasing substi-tutability of raw materials	Allows use of cheaper raw material base
Energy consumption	Less moving parts reduce friction and need for cooling	Information trans-mitance is substituted for the transport of goods and people	Work in the dark and in factories without air-conditioning	Allows production at low temperature and low pressure	More efficient use of energy
Pollution	Better moni-toring	Decentralization reduces travelling		Produces less toxic substances; can help clear polluted sites	New catalysts; less smokestack industries; less waste
Rigid production		Easy switch of production sites	Smaller batches; centrepiece of flexible automation	More substitutes for existing processes; smaller scale of production (bio-reactors)	Larger number of materials; input with flexible automation (light materials)

1.3.6. Core technologies provide more flexibility

The final bottleneck of postwar accumulation had been that it had led
to increasing inflexibilities as the result of efforts to achieve maximum
economies of scale. This trend towards ever larger production runs,
factories and companies may have become reversed by the introduction
of new technologies:

(a) The application of microelectronics and reprogrammable equipment
    has made *flexible automation* possible which allows *small batch
    production* at a price per unit of output comparable to that in mass
    production.
(b) This has led to a considerable reduction of the minimal production
    volume that has to be reached in order to work profitably which makes
    a *smaller average plant size* possible.
(c) Modern telecommunications facilities allow quick reactions to a
    changing environment.
(d) CAD/CAM equipment reduces the overall development time of
    products. It also makes it possible to offer a broader assortment of
    similar products and a better adaptation to specific consumer demands.
    Integration of CAD/CAM equipment with numerically controlled
    machines in manufacturing reduces lead time even further.
(e) Modern communication and data-handling equipment provide central
    management with instant information about every aspect of business.
    This enables management to intervene directly where necessary to avoid
    unwanted developments.[43]
(f) Biotechnology and the development of new materials allow flexible
    changes in the raw materials base because of an increasing
    substitutability of different raw materials.

The different ways in which core technologies contribute to overcome the
bottlenecks of growth, created by the forms of postwar accumulation, are
summarized in Table 1.2. The present chapter has shown that the
historically unique period of growth after the Second World War led to
a number of problems. Many hope the application of new technologies
will help to find a way out of the economic crisis and open up new avenues
for investment and growth.

  A number of core technologies have been defined which are closely
interrelated. These technologies lead to the development of a host of new
products and seem to offer a solution for the problems described. It is

---

[43] Better management control, however, can also imply *less* flexibility, as pointed out in
Chapter Four.

no wonder that European (like American and Japanese) companies throw themselves on these technologies in order to participate to the maximum from the new business opportunities and lay the basis for a new wave of growth. In the next chapter, the activities of a number of European multinationals in the fields identified above will be described.

Chapter Two

Corporate Strategies Towards Core Technologies

European multinationals have taken on the challenge of new technologies. In this chapter, we describe their shift towards 'core technologies'.

Rapidly developing technologies, shorter life-cycles, and increasing R&D expenditures by their competitors confront firms with a number of strategic choices:

—Which are the *prime products* the company will be able to develop, produce and market in the coming decades?
—Which products are an important or even *indispensable input* for these end-products?
—Which activities can easily be *subcontracted* to reliable suppliers?
—Which products can be bought *in the open market* without running the risk of market rigidities?
—Which *production technologies* will the company have to master in order to produce the products of the future at a competitive price?
—Which products can be left aside without damaging the profitability and reputation of the company?

Companies find different answers to these questions, depending on their traditional product range, market circumstances (oligopolistic or competitive markets) and their vision of the company's future. Some companies have made explicit choices to boost their activities in core technologies, others have not. It is not a question of belonging to a specific branch. As we tried to point out in the first chapter, core technologies cut across almost all branches and sectors. It is therefore hardly possible to delineate specific 'high technology *sectors*'. There can be a rather traditional attitude in companies which form part of branches that are regarded as 'high technology' (e.g. telecommunications or aerospace), and there can be a strong orientation towards new technologies in companies that belong to branches which are considered as 'traditional'. To substantiate this point we start with the example of the automobile industry. Later on in this chapter, we shall see how far the reaction of European car producers to the economic crisis of the 1970s can be generalized and whether a similar pattern can be observed at other European multinationals.

Example: multinationals in the automobile industry

The automobile industry is the sector that has been at the centre of postwar economic growth. It produced a durable consumer good which has been crucial for the development of specific production technologies, of the pattern of urbanization and of the pattern of consumption in general. It even gave the name ('Fordism') which is used by many scholars to characterize the postwar model of accumulation.[1]

Most car producers were hit hard by the recession in the 1970s. Many of them showed considerable losses in 1974 and again in the early 1980s. Having been at the core of the 'old' regime of accumulation, the automobile companies were pressed hard to find an answer to the crisis, and part of their answer has been the diversification into the field of 'high tech'.

It is interesting to see that not all automobile producers got into difficulties during the 1970s and early 1980s. Net profits almost tripled at DAIMLER–BENZ AG between 1970 and 1982, and grew by more than 500 per cent at BMW. That means that those companies specializing in the high-value end of the market with a very broad range of models offered to high-income customers, did not experience much of a crisis. This was a signal to other automobile producers that mass production of large series may have come to an end, while smaller series of cars especially suited for specific market segments continue to take a fair chance. The resulting shifts of the range of products towards the high-income market put the successful producers under pressure to perfect their own production techniques as well.

The best-known example of the diversification into the field of 'high tech' is the move of the world's largest automobile producer, GENERAL MOTORS, when the company bought HUGHES AIRCRAFT and ELECTRONIC DATA SYSTEMS (EDS) and formed a joint venture with the Japanese company FANUC, GENERAL MOTORS FANUC Inc., which rapidly became the first American robot producer.

A comparable strategy has been adopted by DAIMLER–BENZ which a few years later (1985/6) bought the engine producer MTU, the aircraft builder DORNIER and finally the troubled electronics firm AEG. These moves are evidence of the company's wish to diversify into areas of electronics not so much to broaden its product range as to control some of the strategic inputs for its still very profitable core business (production of luxury cars).[2] AEG's headstart in radar technology may lead to devices

---

[1] See Aglietta (1976); Sable (1982); Roobeek (1987b); Tulder (1987).
[2] The company's chairman Breitschwerdt 'argues that he must have a solid base in high technology in order to stay competitive in the car market. By sharing the cost of basic research with its new subsidiaries, Daimler–Benz can match the deep pockets of GM and other big competitors'. *Fortune*, 15 September 1986, p. 49.

that detect obstacles in a car's path that a driver does not see. DORNIER is a leader in materials technology. It has developed tough, lightweight carbon fibers and synthetic alloys that may one day be used to build car frames and body panels that do not corrode. DORNIER's navigation and tracking system could be applied to steer themselves.[3] MTU is working on ceramic engine parts. AEG already supplies DAIMLER with the software that commands some of DAIMLER's 790 robots. MTU has been acquired to get hold of its production systems capability. Within a year these acquisitions have brought DAIMLER–BENZ from the periphery of the micro-electronics cluster to its centre.

The recent move of DAIMLER–BENZ into electronics is of a fundamentally different nature from the takeover of the office equipment producer TRIUMPH–ADLER by VOLKSWAGEN in 1979. A couple of years ago, the automobile industry was often regarded as a declining industry, and the investment in TRIUMPH–ADLER was part of a strategy to diversify *out of the automobile industry* into an area (electronics/office equipment) with a bright future (though it has also been the intention to use the knowhow at TRIUMPH–ADLER for an improvement in VOLKSWAGEN's main product).[4] Given TRIUMPH–ADLER's old-fashioned product line, however, VOLKSWAGEN did not find much of a pool of electronics knowledge to tap for the growing electronics content of motor vehicles. Since 1979, it has become obvious that the automobile industry is not a declining sector at all,[5] and VOLKSWAGEN itself has managed to get back to the technological forefront again. As a consequence, diversification out of this industry is no longer a major objective. Consequently, the (still loss-making) electronics subsidiary was sold to OLIVETTI in 1986 in exchange for a 5 per cent share for VW in OLIVETTI.

The different moves — one automobile producer entering electronics, while the other sells its electronics subsidiary — therefore are not an expression of contradictory trends. Both reflect a concentration of companies on core technologies. While the acquisition of AEG, MTU and DORNIER can help DAIMLER to strengthen its high technology orientation, TRIUMPH–ADLER was not of much use to VOLKSWAGEN in this respect. This illustrates very well that 'core technologies' cannot easily be identified with specific branches. Not all electronics companies are 'high tech' and busy in the field of core technologies, whereas parts of the automobile industry

---

[3] *Fortune*, 15 September 1986, pp. 49–50.
[4] *International Management*, October 1982, p. 21.
[5] The high-tech car hits the road, *Fortune*, 29 April 1985, pp. 114–21.

(e.g. VOLKSWAGEN's highly automated production 'Hall 54') may well belong to this field.

While VOLKSWAGEN itself has become active in robot production, DAIMLER–BENZ has not. Most European car manufacturers that started to produce robots for their own use came from the low-value end of the market. The reason for this is that when the first robots were developed, they were not yet very flexible, and therefore were means of mass production rather than of small batches. FIAT, RENAULT and VOLKSWAGEN even became the largest robot producers in their respective home markets. Only VOLKSWAGEN produces robots almost exclusively for its own needs (see 2.4).

All major European manufacturers have been diversifying into the area of *new materials*. They have adopted changing production methods in which more use is made of light synthetic materials (plastics) and less of metals.[6] Car producers are also very active in the development of energy extensive engines using ceramics. Most car producers have built their own research and development capability in new materials in order to master the technology for strategic inputs. Cooperation among car manufacturers is widespread in many areas[7] and a couple of European car manufacturers have started to cooperate in pre-competitive research in new materials such as ceramics and synthetic fibres. In 1985 FIAT–LANCIA, BRITISH LEYLAND, VOLVO, VOLKSWAGEN, RENAULT and PEUGEOT decided to start a permanent exchange of information and ideas on research but also on the production of these new materials.[8]

In general, thus, the European car industry has diversified into core technologies with a special emphasis on factory automation. The remainder of this chapter will consider the diversification and specialization strategies of major European multinationals in the six core technologies in the microelectronics cluster (sections 2.1–2.6), in biotechnology (2.7) and in new materials (2.8). This has resulted in an increased importance of R&D in company's strategies (2.9). The chapter will be summarized by giving an overview of the activities of a sample of 41 European multinationals studied throughout this book and by considering whether a certain division of labour is taking place among them (2.10). It will be asked to what extent firms have chosen to develop the technologies only for their own use (*captive production*), predominantly for sales (*merchant production*) or for a combination of both (*captive–merchant production*). It can be expected that the resulting pattern of specialization, diversification and

---

[6] Gonzales-Virgil (1985) provides an example of the Ford Motor Company; p. 24.
[7] See Altshuler *et al.* (1984).
[8] *Technieuws*, a publication of the Dutch Ministry of Economic Affairs, Brussels, June 1985.

cooperation will lead to a higher level of concentration in the production of core technologies not only in Europe but also worldwide.

2.1. SEMICONDUCTORS: THE CHOICE BETWEEN STANDARD AND CUSTOM

Semiconductors is undoubtedly one of the most prominent areas in which not only commercial but also strategical considerations play a vital role. Semiconductors are the 'core of the core' of the microelectronics cluster and are vital inputs in many end-products. This has become more relevant since the semiconductor content of telecommunications equipment, computers, consumer electronics and industrial equipment has rapidly increased.

The semiconductor industry is highly volatile with a rapid succession of large overcapacity and extreme shortages. This is especially true for the standard chip market, which as a consequence is extremely risky. The basic strategical problem for firms wanting to have their own semiconductor capability, therefore, becomes which end of the market they have to emphasize: standardized chips which can be sold on the open market or (semi)customized chips which can only be used in specific end-products.

At the end of the 1960s and the beginning of the 1970s clear decisions were made by most European electronics firms. A radical example is provided by GEC which in 1969 decided to depart almost totally from the production of semiconductors, arguing that it did not need to make the devices if they could be bought cheap from other sources. GEC instead wanted to concentrate on integrating the chips into its own products.[9] PHILIPS, SIEMENS, AEG–TELEFUNKEN and THOMSON–CSF with their more diversified product range in consumer electronics, telecommunications and industrial applications did keep a large presence in many groups of semiconductor devices, both standardized and (semi)custom. The four firms operated as captive–merchant producers with a considerable share of the chips integrated into their own end-products. But in the first half of the 1970s they did not move into mass production of Large Scale Integrated (LSI) devices, as did the American merchant producers, probably because their end-products did not require this input yet. The firms with large production for military needs in France and Britain (MATRA, PLESSEY, FERRANTI) committed themselves to a specialization (niche)

[9] McLean and Rowland (1985), p. 28.

strategy and concentrated on customized chips for military equipment. The spin-off for civil business has been extremely meagre since the quality standards of military production and the low batch volumes require totally different knowhow, production competence and company culture. The cost-plus basis on which the defence departments generally bought the devices nevertheless provided the firms with solid profit margins. In an early phase of the semiconductor industry these companies thus specialized into niches and retreated from the standard chip business.[10] This strategy of customized and predominantly captive production becomes difficult to reverse in a later state of the industry's development.

In Italy the small independent chip-producing firm SGS was purchased by the STET group in 1971. This made a concentration of the national semiconductor industry possible (SGS and ATES, the original STET chip subsidiary, merged). The demand for semiconductors from within the STET group was not large, however. This retarded an effective integration of the semiconductor business with STET's other activities and induced SGS–ATES to operate in the open market and, given its limited resources, to choose a specialization strategy. As a consequence the French, British and Italian markets in the first half of the 1970s were already dominated by non-resident producers with the major national firms only producing for market niches.[11]

Since the beginning of the 1970s the whole microelectronics cluster has undergone drastic change because of the development of LSI circuits and of microprocessors in the United States. Periodical shortages in the supply of standardized LSI components from the merchant American suppliers (often discriminating against European firms) additionally induced most European electronics firms to adopt a more active stance. Many large European and American firms acquired (part of) specialized American merchant chip producers, sought to conclude cooperation deals or applied for second sourcing agreements[12] with leading US producers. The resulting wave of acquisitions in the second half of the 1970s changed the whole outlook of the chip industry radically. Originally consisting of a relatively large number of medium sized independent producers, the strategies of the large firms made the semiconductor industry become dominated by large vertically integrated companies (see Table 2.1).

---

[10] Even where they had established real technical advantages in the manufacture of some early standard semiconductors. McLean and Rowland (1985), p. 19.

[11] See Malerba (1985), pp. 151–63.

[12] 'Second sourcing' was originally used in the defence sector where the user of the device wanted to avoid dependence on a single supplier. In semiconductors, particularly microprocessors, second sourcing implies that the original manufacturer agrees to transfer the design and often the production masks to another firm in order to allow the recipient to make the same microprocessor in his own factory.

Table 2.1   The process of vertical integration with worldwide IC producers

1972	1977	1987
AMD	AMD	AMD
AMI	AMI	Intel
AMS	Electronic Arrays	National Semi
Electronic Arrays	Fairchild	
Fairchild	Intel	
Intel	Intersil	
Intersil	MMI	
MMI	Mostek	
Mostek	National Semi	
National Semi	SGS–ATES	
PMI	Siliconix	
SGS–ATES	Synertek	
Siliconix	Zilog	

Main-line merchant IC producers (worldwide)

1970	1976	1987
ITT	AEG TELEFUNKEN	Bourns/PMI
Motorola	Ferranti	Exxon/Zilog
Raytheon	GI	Ferranti
RCA	Harris	Fujitsu
TI	Hughes	GE/Intersil
	ITT	GI
	Motorola	Gould/AMI
	NEC	Harris
	PHILIPS/	Hitachi
	SIGNETICS	Hughes
	PLESSEY	ITT
	Raytheon	Matsushita
	RCA	Mitsubishi
	Rockwell	Motorola
	Sescosem	NEC
	SIEMENS	OKI
	TI	PHILIPS/SIGNETICS
		PLESSEY
		Raytheon
		RCA
		Rockwell
		Schlumberger/Fairchild
		Sharp
		SIEMENS
		Sprague
		THOMSON CSF/
		MOSTEK/SGS–ATES
		Thorn–EMI/Inmos
		TI
		Toshiba
		TRW

Vertically integrated/merchant IC vendors (worldwide)

Sources: Mackintosh (1986), pp. 120, 121; own observations.

In Europe, SIEMENS and PHILIPS in particular adopted an aggressive acquisitions strategy. SIEMENS acquired a number of smaller US companies: Dickson (1974), Litronix (1977), Microwave Semiconductor (1979), Datbit (1979) and Threshold Technology (1980) were taken over for 100 per cent, while SIEMENS took a 20 per cent share in Advanced Micro Devices (AMD), one of the larger specialized semiconductor companies. PHILIPS acquired Amperex in 1972 and one of the market leaders, Signetics, in 1975. Firms like BOSCH, FERRANTI, CIT–ALCATEL, too sought American acquisitions in order to diversify upstream into LSI chip technology.

This was also the case with SCHLUMBERGER, a large French oil equipment firm. In 1979 it acquired the once very innovative FAIRCHILD company. Interestingly enough, this move blocked the strategy of GEC which had established a joint venture with FAIRCHILD in 1978 in order to re-enter the standard integrated circuits market having reappraised the strategic value of these devices for its own business. The joint venture of GEC and FAIRCHILD was consequently terminated in 1980.

The lack of an indigenous standard chip producer induced the British Labour Government to finance a start-up firm in memory chips and microprocessors, INMOS, which was expected to have important spin-off effects for the whole British electronics industry. The move was much criticized by the leading British suppliers of custom chips (FERRANTI, PLESSEY and GEC), fearing that they would lose scarce public funds. When a few years later the Conservative Government decided to privatize the loss-making firms, GEC, however, was one of the candidates to acquire INMOS. Eventually, the firm was taken over by THORN–EMI which had also become convinced of the usefulness of having its own (V)LSI chips and microprocessor facility. Ironically, the latter thought this to be especially helpful in its strategy to move from consumer electronics with its low profit margins into the military business[13] which had been exactly the opposite of the original intentions of the British Government for INMOS. Since the integration of INMOS into THORN–EMI proceeds slowly, INMOS will remain one of the rare European semiconductor companies with a predominantly merchant chip production in the 1980s.

In the second half of the 1970s, in sum, most European consumer electronics, computer and telecommunications firms made extensive efforts to guarantee themselves a strategic in-house capability in the latest generation of semiconductors. A notable exception in this pattern is NIXDORF which has chosen not to dedicate its (limited) research funds to semiconductor technology, but to rely on either cooperation with some smaller firms or on buying in the open market.

---

[13] See McLean and Rowland (1985).

Also with regard to *microprocessors*, most European firms have not been at the forefront of developments. In 1980, only PHILIPS ranked among the first ten producers of microprocessors worldwide.[14] Instead of developing microprocessors themselves, they heavily relied on second sourcing agreements with important American producers. For the large firms *and* for the smaller innovator, a strategy of second sourcing became very viable. For the original manufacturer it became possible to *set a widely accepted standard* by allowing large firms to produce the same device, to *gain rapid entrance* to markets beyond their reach and to team up with large firms which can *develop additional devices*, applications and supportive chips which make the processor more suitable for various functions. For the large firms at the receiving end of the technology, second sourcing enabled them to *follow the latest technological innovations*, without running the risk of large R&D costs for eventually unsuccessful devices. Since some second sourcing agreements include far-reaching cooperation, it gives both firms a larger capacity in the development of new generations of chips. SIEMENS, for instance, helped INTEL directly with the development of its 32-bit chip as part of its second sourcing agreement with INTEL.[15] In microprocessors, therefore, the cooperation between large European firms and smaller American firms often is on an equal basis, guaranteeing that the components of both firms are completely compatible.[16] Second sourcing enables the larger firms to *wait for the standard* version in the newly developed generation of microprocessors. The relative success of SGS in 8-bit microprocessors, for instance, was partly due to the timely licensing of the Z-80 microprocessor from Zilog. PHILIPS originally chose the 8-bit processor of INTEL but switched to the competing standard in 16-bit microprocessors of Motorola. Second sourcing thus is an attractive alternative to direct acquisition of the innovative company in question (with all the problems of keeping the skilled people in the firm and bearing all other commercial risks). There are, however, considerable risks involved in second sourcing as will be pointed out with the example of THOMSON later on.

With the coming of Very Large Scale Integration (VLSI) in the 1980s[17], the degree of worldwide concentration in the standardized semiconductor and microprocessor markets is bound to increase for obvious reasons: higher development expenditures, closer ties between users and producers and the necessity of large minimum sales to get enough return on

---

[14] i.e. it was tenth! Lorenz (1982).

[15] *Business Week*, 6 August 1984.

[16] Siemens press release, The Hague, 23 September 1982.

[17] 'This period is characterized by semiconductor devices with sub-micron channel width, by microprocessors of 32-bit units or more, and by dynamic RAM memories of 256K of information or more'. Malerba (1985), p. 241.

investments in production plants and R&D. The R&D requirements for chips as a percentage of total sales have risen from around 10 per cent in 1978 to more than 25 per cent in the 1980s. Additionally, the minimum investment for setting up a semiconductor production line increased from $100 000 in 1954[18] to more than $100 million in 1984. It is therefore not sufficient to acquire one leading semiconductor company. Firms are forced to continue their R&D efforts with increased speed, increased investments and great uncertainties whether the end-product can be commercialized successfully.

This is not only necessary for retaining their own capability in chips, but also to remain an attractive and powerful *potential partner* in cooperation and second sourcing deals with American and Japanese producers which are vital for the more diversified European producers since it is impossible to achieve a competitive position in *all* areas of semiconductors independently. Second sourcing agreements in the next generation of 32-bit microprocessors, however, have not been as easy to achieve as previously, especially for firms not able to offer an equivalent product in return for becoming the second sources and due to more stringent export control measures by the American Government. THOMSON, for instance, experienced great difficulties in extending its agreement with MOTOROLA on 16-bit processors into the 32-bit area. The chip technology in telecommunications applications which THOMSON could deliver as part of the deal was not sufficient as a 'bargaining chip' according to MOTOROLA.[19] The uncertainty surrounding the negotiations has severely handicapped THOMSON's competitiveness in other areas as well. Giving up the capability in some parts of semiconductor R&D and production might also jeopardize the firm's autonomous position in other areas due to the loss of knowhow needed in converging areas and advanced applications. Another example is ERICSSON which 'started buying integrated circuits from foreign manufacturers when its own component subsidiary, RIFA AB, could no longer match demand. Now it is buying processors too'.[20] ERICSSON (like other large European firms) had to use all its combined buying power in 1984 to assure deliveries of vital components from its American suppliers when these merchant producers could not match worldwide demand.[21] It might be a prelude to more difficulties once the firm becomes too dependent on outside deliveries of chips and especially of the latest generations of integrated circuits (memory chips and microprocessors). This can severely hamper its competitive strength in telecommunications.

---

[18] Truel (1980).
[19] *Computable*, 22 August 1986.
[20] *International Management*, October 1985, p. 32.
[21] Interviews.

This is the main reason why PHILIPS, SIEMENS and also THOMSON invest huge amounts in the next generation of memory chips—not to become the first company in commercializing this (submicron) technology, but to have an in-house capability in one of the most strategic areas in semiconductors. SIEMENS in its 1984 annual report explicitly stated: 'we intend to make ourselves less and less dependent on outside suppliers for the semiconductor circuits used in our own systems, while simultaneously expanding our components business on the world market'.[22] SIEMENS and PHILIPS, subsidized by the German and Dutch Governments, joined forces in the Megachip area in a Japanese style in which the two firms share common R&D and production knowledge, but use it in different areas (SIEMENS in dynamic Random Access Memories (RAMs) and PHILIPS in static RAMs). Both firms have to build entire new facilities for the Megachip project involving well over DM500 million of investment for each firm.

THOMSON did not participate in this cooperation, but THOMSON's close links with the military make a domestic capability in components equally mandatory. About 35 per cent of R&D in components with THOMSON is paid for by the military. Still this is not enough to ensure a satisfactory domestic production capability in semiconductors. The heavy dependence on US components is regarded as unsatisfactory. 'THOMSON cannot afford not to control its semiconductor technology.'[23] In 1986 this induced the firm, for example, to acquire the bankrupt integrated circuits producer MOSTEK from UNITED TECHNOLOGIES despite the fact that the firm not only was loss making but also had missed the step towards the latest generation of (256K) memory chips. Additionally THOMSON received FF1 billion from the French Government to develop its own megabit chip. This move probably has spurred negotiations between the three European firms determined to participate in the worldwide race in standardized chips on cooperation for the next generation of standard memory chips (16, 32 or 64 megabit).

The only other European producers which might be able to jump on this bandwagon are SGS–ATES and perhaps BRITISH TELECOM. The STET subsidiary also has kept a capability in (V)LSI chips and is the fourth largest European chip producer after PHILIPS, SIEMENS and THOMSON. In 1987, the concentration level in the European chip industry was considerably raised by the merge of THOMSON's chip department and SGS. BRITISH TELECOM provides an example of a late-comer in the chip business, because it had no in-house capability in chips before

[22] Siemens, *Annual Report, 1984*, p. 32.
[23] According to Mr Gomez, chairman of Thomson, in *Financial Times*, 25 October 1985, p. 14.

privatization. 'As a public corporation, BT was happy to contract out its chip production and some research, but as a privatized company BT management now wants an internal source of integrated circuits.'[24] By mid 1985 the firm was actively trying to acquire a chip-producing firm and already spending £20 million to £30 million per year in semiconductor research, which is approximately 15 per cent of its total annual R&D budget. Compared to the large expenditures of firms like SIEMENS, PHILIPS or THOMSON the amount is not yet very large and will not make BT a prominent force in semiconductors. BRITISH TELECOM probably is one of the last European electronics firms which has integrated backwards into chip production. All other new activities are dominated by the already established firms.

Whereas in the USA around 50 chip start-ups were founded in the 1978–85 period, only four start-ups can be counted in Europe.[25] These start-ups were all either initiated by large European multinationals, state supported, or a combination of both. The four firms are:

(a) INMOS; in 1978 set up by the British Government and in 1984 taken over by THORN–EMI.
(b) MATRA–HARRIS, the 1978 joint venture of MATRA and the US firm HARRIS.
(c) MIETEK, a joint venture of the Flemish Government and ITT–Belgium.
(d) EUROPEAN SILICON STRUCTURES (ES2), initiated in September 1985, which specializes in (semi)custom chip design and is financed with some venture capital and by seven large European shareholders: PHILIPS, BULL, SAAB–SCANIA, OLIVETTI, BROWN BOVERI, TELEFONICA and BRITISH AEROSPACE.[26]

The last and most recent project points at a new phase in the strategies of major European firms, i.e. to dedicate more resources to the production of customized chips. A number of firms have concluded cooperation agreements for this goal (see Chapter Seven). PHILIPS and TEXAS INSTRUMENTS, and SIEMENS and TOSHIBA have combined their design resources in databanks from which customized chips serving captive as well as merchant ends can easily be generated. This corroborates the general strategy of the big European electronics firms to target the high value added end of the market and become more service oriented (see

---

[24] Bill Jones, chief executive of Technology with BT, *Electronic Times*, 13 June 1985.
[25] See Jean-Claude Reflet, director of Matra–Harris, quoted in *Electronics Weekly*, 9 May 1985.
[26] The seven participating firms have each less than 5 per cent of the shares. They raised a total of $36 million. Together with loans of around $25 million the company will have raised $61 million. ES2 is also one of the approved EUREKA projects (see Chapter Seven).

Chapter Three). But since firms like PHILIPS and SIEMENS want to keep their highly diversified nature, they cannot confine their resources to customized chips only, and need to stay in standard chips as well.

2.2. COMPUTERS: THE CONTINUOUS BATTLE WITH IBM

The weakness of most European firms in semiconductors has partly been due to their inability in developing an adequate strategy in computers. Particularly in the USA the 'demand pull' effect of the computer industry has been far larger than in Europe (see Chapter Five). On the other hand, semiconductors in general and (since 1971) microprocessors specifically acted as 'technology push' factors for new computer generations and applications.

Since the 1960s it has not been possible to analyse any company strategy in computers without reference to the world market leader, IBM. In 1985 total world sales of IBM were more than six times the turnover of its best-placed competitor (Digital Equipment Company). Even when two of the five largest data-processing firms in the world (Sperry and Burroughs) merged in 1986, their combined sales reached only 20 per cent of IBM's turnover. Among the world's top ten computer firms is only one European firm, SIEMENS, which ranks ninth. SIEMENS receives 85 per cent of its computer income from sales in Europe (particularly in Germany). This is a common pattern for other European 'national champions' in computers such as ICL in Britain and BULL in France. Due to preferential procurement and other support by their governments they managed to stay in the race, with most of their sales in the home market. Preferential public procurement in France, for instance, had the effect that French manufacturers (i.e. mainly BULL) provided 63 per cent of the civil service's computer installations, whereas they only served 45 per cent of the private market.[27]

The status as a national champion in computers induced these European firms to pursue a strategy of producing a 'full product line' based on proprietary operating systems.[28] The competition with IBM over the whole range of computer systems has not been very effective because it dispersed limited funds over too many applications which only could be sold on the limited national markets. The European turnover of IBM is

[27] US Department of Commerce (1983b), p. 32.
[28] Mackenzie (1985), p. 14.

Table 2.2   Top ten data-processing companies* in Europe, 1985

Rank 1985	Rank 1984	Rank 1981		Headquarters	1985 European Sales ($m)	% of Euro market
1	1	1	IBM	USA	13 440	33.5
2	2	3	SIEMENS	Germany	2 755	6.9
3	4	4	DIGITAL EQUIPMENT	USA	2 179	5.4
4	3	6	OLIVETTI	Italy	1 863	4.6
5	5	2	BULL	France	1 669	4.2
6	8	12	NIXDORF	Germany	1 193	3.0
7	9	10	BURROUGHS	USA	1 124	2.8
8	6	—	ERICSSON	Sweden	1 097	2.7
9	11	9	PHILIPS	Netherlands	1 078	2.7
10	10	5	STC–ICL	UK	1 038	2.6

Sources: Datamation (1986); own calculations
*including computer hardware, software and data communications equipment (therefore also ERICSSON included).

only 28 per cent of its worldwide turnover, but still is almost five times SIEMENS's European computer sales (see Table 2.2).

The only firms with considerable success in computers have developed a strategy in *specific* market segments which (initially) were left aside by IBM long enough to let these firms establish a foothold in these markets.[29] This happened to some extent with OLIVETTI and NIXDORF in the minicomputer market. In the traditional core business of IBM, mainframes, PHILIPS, SIEMENS and BULL tried to develop their own commercial models in the UNIDATA project (1972–5). The project failed due to BULL's preference to cooperate with an American partner (Honeywell). The failure of UNIDATA prompted PHILIPS to withdraw from this market segment altogether.

The most viable strategy in mainframes since the 1970s has not been to develop an own proprietary computer system, but to make so-called 'IBM plug compatible' equipment. Such computers are based on the IBM standard but are sold at lower prices. In the 1970s this strategy has been pursued by the Japanese late-comers Fujitsu and Hitachi, with considerable success — witness the fact that many European national champions now market Japanese (mainframe) computers. SIEMENS and ICL market Fujitsu mainframes[30] in addition to their own mainframes; BASF and

[29] It has sometimes been suggested that IBM might 'tolerate' these firms in order not to provoke anti-trust measures.

[30] ICL, however, not very successfully. From 1982 to 1984 it could only sell six Fujitsu mainframes, whereas Siemens sold about 200 computers (larger and smaller versions); ICL in 1984 therefore terminated the marketing of the computer but continued the technology transfer agreement with Fujitsu. *Financial Times*, 14 July 1984.

OLIVETTI market Hitachi mainframes without any own capability in this area. For BASF, this agreement has been part of a diversification strategy. It already produced the tapes and diskettes used in computers, but has chosen to move more into the core of the computer business by producing the hardware, partly under licence.[31] In 1987 SIEMENS and BASF decided to join forces and merge their mainframe computer activities. BULL chose to team-up with Nippon Electric Company (NEC) in mainframes, next to a long-term technology exchange with Honeywell. Honeywell (even after the nationalization of the joint venture CII–HONEYWELL BULL in 1982) still had an 8 per cent share in the company. In order to restructure its loss-making activities, BULL is increasingly extending a cooperation strategy. BULL is 'emerging from the "ghetto" of refusing to rely on outsiders in which it had been trapped under previous government policies. "We will never do anything completely on our own" . . .'.[32] This happened even when BULL took over the computer division of Honeywell in 1987, in which also a minority share became held by NEC.

Since the beginning of the 1980s, it has been especially in the intermediary range of *small and medium sized computers* that European multinationals have concentrated their efforts. All mainframe manufacturers have diversified into the production and marketing of smaller computers. The same is true for the *telecommunications firms* (ERICSSON, PLESSEY, GEC, PHILIPS, CGE) which need computer technology as a strategic input for their own digitalized end-products. ERICSSON, for instance, took over the Swedish computer firm FACIT and is now also marketing business computers and personal computers (PCs). Its public exchange system, AXE, is mainly based on medium sized computers. The same strategic considerations in 1984 induced the telecommunications company STC, formerly a British ITT subsidiary, to acquire ICL.

Another broad category of firms diversifying into the area of personal and business computers consists of former specialized *typewriter producers*. The main reason here is the increasing substitution of their old core activities by digitalized devices. OLYMPIA (AEG), TRIUMPH–ADLER, and OLIVETTI itself are major examples. These firms, however, traditionally have not been very R&D intensive. OLIVETTI, for instance, spent only around 3 per cent of its sales on R&D in 1975 before Mr De Benedetti became chairman. In absolute figures this was only 1.5 per cent of the R&D budget of PHILIPS which in 1975 spent more than 7 per cent of its (far larger) turnover on R&D. Both German

---

[31] Computer activities of BASF in 1985 accounted for a turnover of more than DM600 million. In 1986 negotiations with Siemens started on a possible merger of their mainframe business which were finalized in 1987.

[32] Jacques Stern, president of Bull, in *Financial Times*, 20 March 1984.

typewriter firms consequently teamed up with capital-rich firms, whereas OLIVETTI after 1978, eager to become a major force in the 'Office-of-the-future', adopted a strategy of extensive cooperation with other firms. Since 1980 OLIVETTI acquired shares in many small high tech American companies in computer and software technology.[33] The reason behind these acquisitions is to have access to advanced technology through participation, rather than winning important shares of the US market, since these small firms often had only one product, or had not even reached the stage of commercialization. OLIVETTI treats these new companies as independent units, in order to retain their 'entrepreneurial spirit'.

In 1984 OLIVETTI struck its major deal by teaming up with the American telecommunications giant, AT&T. In exchange for a 25 per cent share (which can be extended to a maximum of 40 per cent) in OLIVETTI, AT&T offers OLIVETTI access to its massive knowhow in telecommunications and will market OLIVETTI computers in the USA. Since then it seems that the participation and cooperation strategy of OLIVETTI has tended towards Europe, where it acquired ACORN computers in the United Kingdom and TRIUMPH–ADLER in West Germany (among others). This strategy makes a rapid extension of the marketing network of OLIVETTI in Europe possible.

The strategy of OLIVETTI with regard to the transfer of technology through (American) minority participation resembles that of NIXDORF, the only other comparatively small, successful European data-processing firm which is not a national champion. NIXDORF has also acquired many minority shares in small high-tech firms in the USA, but has relied much more on its own strength in R&D which consequently accounts for almost 10 per cent of its total turnover.

Despite the fact that OLIVETTI and NIXDORF might be considered newcomers in computers, the rate of concentration in Europe has not altered drastically, with 55 per cent of total sales in the hands of only five producers and around 70 per cent with the first ten firms. Table 2.2 shows that in 1985 seven European firms rank among the first ten producers in Europe. This picture is misleading, however, since the combined sales of US-based companies in Europe nevertheless almost doubled the European companies' sales. Twelve of the first twenty-five companies in Europe are of US origin.

At the *users'* end, probably the majority of European multinationals (especially in the area of office automation) have bought or leased IBM mainframes. A well-known dictum with procurement managers in large

---

[33] In the 1980–4 period, for instance, Olivetti took a minority share ranging from 2 per cent to 46 per cent in at least 24 firms in the USA. See Wiltgen (1987), Chapter II; Ciborra (1986).

firms is 'no one has ever been fired for buying an IBM computer'! The resulting close user–supplier relation with the large firms has given IBM a place in the very heart of these firms which additionally have come to rely on IBM peripherals and other services as well.[34]

Until the end of the 1970s, computer systems were often so-called 'stand alone' models, i.e. not linked to other computer systems. With the transmission of data becoming inseparable from data processing, the boundaries between the computer sector and telecommunications have become vague.

2.3. TELECOMMUNICATIONS EQUIPMENT: MERGING TO STAY IN THE RACE

The international telecommunications market is highly concentrated. The three largest manufacturers (Western Electric (AT&T), SIEMENS and CGE–ITT) together take care of more than 40 per cent of the world market, whereas the ten largest firms account for more than 70 per cent.

Of the largest firms only Western Electric, ERICSSON and Northern Telecom have more than half of their sales in telecommunications equipment. This implies that most firms operating in the market already are very diversified companies.

The (state) monopoly in postal and telephone services in most countries created a monopsonistic market structure in which it has been very advantageous and often a prerequisite for getting the orders for the suppliers of equipment to cooperate with the national service carriers. Close ties between users and producers, therefore, have stimulated the concentration process in the telecommunications industry resulting, in most countries, in a share of the four largest firms in total sales of more than 80 per cent. The largest supplier in most countries is a 'national champion': GEC in the UK, ALCATEL–THOMSON in France, SIEMENS in Germany, STET–ITALTEL in Italy, ERICSSON in Sweden, PHILIPS in the Netherlands and BELL Telephone Company (Alcatel–ITT) in Belgium. With the privatization of BRITISH TELECOM in the UK a new contender has come to the fore which is likely to increasingly plan for and produce its own equipment requirements. In France, the rate of concentration has grown considerably after the nationalizations in the 1970s and in 1982

---

[34] With regard to the Netherlands, for instance, it is observed that if the around 200 IBM maintenance employees were to go on strike, the whole Dutch economic system would crumble.

(especially of the ERICSSON and ITT subsidiaries) and the merging of the telecommunications activities of CGE and THOMSON in 1984.

In many industrialized countries deregulation of part of the PTT's monopoly position is taking place (see Chapter Six). This has introduced greater competition at the lower end of the switching, terminal and transmission markets (Private Business Exchanges (PABX), Local Area Networks (LANs), Value Added Networks (VANs)). Especially in this area, data-processing firms have taken on the challenge and diversified into the production of telecommunications equipment. The difference between a small or medium sized computer and a PABX system (with a few hundred extensions linked to it) is not very large. The most prominent example of an offensive diversification strategy is provided by IBM which in the 1980s rapidly bought its way into the telecommunications business by acquiring ROLM, a medium sized PABX producer, by moving into satellite communications (by founding the firm Satellite Business Systems) and also by acquiring a share in MCI, a major American telecommunications firm, in order to complement its own capabilities. In Europe, OLIVETTI gained a 20 per cent share of the Italian PABX market within five years at the expense of such traditional suppliers as ITALTEL. For this move OLIVETTI relied on licensed foreign technology. The link-up with AT&T might give OLIVETTI even more strength in telecommunications although the main benefit for OLIVETTI clearly is the reasonably strong sales of OLIVETTI-built AT&T personal computers, which has little to do with its position in telecommunications. As in computers NIXDORF has developed a strategy similar to that of OLIVETTI (and IBM) though on a much smaller scale. In 1982, it branched into telephone switching, introducing Germany's first digital system for corporate users even ahead of SIEMENS. NIXDORF has close ties with FERRANTI of the UK and cooperates with Japanese companies for intelligent facsimile. BOSCH and MANNESMANN AG in Germany also invested in the telecommunications equipment industry by taking over part of the telecommunications activities of AEG, mainly represented by its share in Telefonbau und Normalzeit (T&N). NIXDORF, OLIVETTI and BOSCH are becoming major competitors in the European telecommunications industry, diversifying from their traditional product range into this area.

In what might be called the core business of telecommunications, *public switching exchanges* (accounting for around a third of the total sector), a clear concentration of production and development already is taking place. Most European producers try to keep a certain autonomy, but are forced to cooperate more with each other or with their American counterparts. In fact, most European firms are forced to merge their telecommunications activities in order to stay in the intensified race for the next generation of (digitalized) public switching exchanges. Of the approximately twelve firms which had separately developed an important digital switching system, at least ten have joined their activities one way or another in the 1983–6 period! All these agreements include a European partner. PLESSEY, which together with GEC (and STC when it still was an ITT subsidiary) had developed System X, acquired the US company Stromberg–Carlsson. This was inspired by the wish to expand into the US market. It was helped by the serious difficulties of the (small) American firm to get a sufficient rate of return on its initial investment in its costly digital system. A small market penetration and high R&D expenditures for an updated digital exchange system were important reasons for PHILIPS in 1983 to line up with AT&T-International in combining efforts in the joint development of a new digital exchange system. AT&T from its side was not only interested in the cooperation because of high development costs, but also wanted to expand its market potential in Europe through the distribution channels of PHILIPS. PHILIPS and AT&T, however, were unable to negotiate a more ambitious co-development arrangement. The two French firms THOMSON and CIT-Alcatel in 1985 merged their telecommunications business. In Italy ITALTEL and TELETTRA (the FIAT telecommunications subsidiary) in 1986 tried to form a joint venture. In 1986 SIEMENS tried to join forces with GTE, one of the largest American equipment producers. As a result SIEMENS took over most of GTE's European telecommunications business. In 1986 also the most spectacular acquisition took place: ITT sold its entire telecommunications division to CGE. This made CGE at once the second largest producer of telecommunications equipment in the world after AT&T. The deals of SIEMENS and CGE have overshadowed another action undertaken in 1985 by these two firms together with PLESSEY and ITALTEL, but which will also have important effects for the future of European telecommunications: the four firms agreed to form a joint R&D team for the development of components for the next generation of digital exchanges, a clear step towards standardization in this area. In October 1987, PLESSEY and GEC finally announced plans to merge their telecommunications activities in a joint company. In Europe this only leaves ERICSSON as a firm independently developing its own digital switching systems.

2.4. ROBOTICS: TARGET AREA FOR MACHINE TOOL, AUTOMOTIVE AND ELECTRONICS COMPANIES

The newest generations of semiconductors, medium sized computers and telecommunications technologies find an important area of application in the core of production automation, i.e. robotics and other advanced machine tools. Robot production has been an area of activity of machine-building companies, in the first place. Machine building is a sector with many small and medium sized companies with a low level of concentration because the broad spectrum of machinery used in other sectors offers many different fields of specialization for a large number of companies.

Whereas quite a large number of machine-building companies have tried to enter the field of robot production, there has been, right from the start, a rather strong concentration on the side of the *users* of robots. Most of the robots installed up to now are employed in the automobile industry, which is a highly concentrated industry. The automobile firms themselves have the best knowledge of their own production processes, and many of them are used to producing part of their own machinery. Several of the largest European car producers, especially FIAT, VOLKSWAGEN and RENAULT, consequently started to produce their own robots. Since they were able to use a large number of robots in their own factories, they could accumulate considerable experience and were able to realize important economies of scale. Three out of four of the largest European car manufacturers have consequently also become the largest producers of robots in their respective countries. Some 20 per cent of total robot installations in Germany is with VOLKSWAGEN. FIAT, which owns a majority share in robot producer COMAU (another 30 per cent is owned by the US Bendix corporation) has made enormous investments in the production of robots, which are also exported to other car manufacturers. The same is true for RENAULT's subsidiary ACMA which delivers about 50 per cent of its production to other automobile producers including PEUGEOT. Around 30 per cent of all robots installed in France come from ACMA.

In France, Italy and Germany, the major robot producer thus is at the same time the major robot user. Another car manufacturer which takes care of part of its own robot needs is VOLVO which has produced several hundred units of special-purpose industrial robots[35] besides buying from outside suppliers. In Germany BMW did not produce robots, but BMW engineers actively participated in the development of industrial robots with its national supplier (KUKA), as the BMW part of the Quandt Group. KUKA is one of the most successful 'merchant' European robot producers, but probably only due to this close relation with BMW. VOLKSWAGEN, DAIMLER–BENZ, BMW and SIEMENS as well as the City Government of West Berlin formed a joint venture in Berlin for research on robot programming, sensor technology, flexible automation and data systems

[35] Carlsson (1983), p. 23.

in automobile production.[36] It is one of the examples of an effort to follow the Japanese example of close intercompany cooperation in basic research.

Another group of users that becomes more important is the *electronics industry*. With robots becoming more sophisticated they have become able to perform assembly tasks. This has made them very attractive for electronics firms, which have a very high labour cost component,[37] mainly because of the asembly work. In Japan, robots used in the electrical machinery industry have already outnumbered those used in car manufacturing. A similar development is taking place in Europe.

Being themselves an important (potential) user of robots is not the main reason why electronics firms are interested in robot production. They are increasingly important in the robot industry because the electronic component of robots has become ever more crucial to the development of sophisticated and 'intelligent' robots of the third generation. With the electronic components being responsible for more and more of the value added in robot production, the electrical engineering companies promise to have a leading edge in the future. This process is especially apparent in the USA and Japan where large electronics companies have diversified into the area of robot development and production.

In Europe, developments have been different in the past. In the field of electronics, the large European companies have not been at the forefront of international technological development during the last decade. Consequently, the largest European firms in the field of electrical machinery do not have the same prominent position as companies in the same sector in the United States and Japan. European machine-building companies (and car producers) are more competitive, internationally, and robot production in Europe has been more the realm of other companies than those in the field of electrical machinery.

None the less, all major European electronics firms have a capability in robot technology. PHILIPS developed some robots for its own use but has not offered its two prototypes of a second generation assembly robot in the market.[38] SIEMENS has founded the company Mantec GmbH which has a merchant strategy. SIEMENS cooperates with Fujitsu Fanuc (in which it still has a 12 per cent share) in robot development and markets the Japanese robots on the European market. Another German electronics company entering the robot market in 1981 as an integral part of its own diversification strategy is ROBERT BOSCH GmbH. In the United Kingdom GEC became the largest indigenous robot producer when the

[36] *Frankfurter Allgemeine* (Frankfurt), 16 June 1983, and other sources.
[37] *Journal of Commerce* (New York), 25 July 1983.
[38] Philips, *Technisch Tijdschrift*, vol. 1981/2, no. 40.

company bought Hall Automation in 1979.[39] But since the robots developed by Hall Automation proved to be outdated, GEC signed an agreement with HITACHI at the end of 1982 to market and eventually produce HITACHI robots under licence for the British market.

In Italy, OLIVETTI is the second largest of Italy's robotics companies with sales only slightly behind the FIAT subsidiary COMAU. The robot subsidiary of OLIVETTI, OSAI, in 1982 signed an agreement with the American company Allen–Bradley, a prominent firm in production automation systems which SIEMENS unsuccessfully tried to acquire in 1985. This led to the formation of a new company, OSAI A-B (68 per cent owned by OLIVETTI), which also operates in the UK, Germany and France. In 1981 STET took over Digital Electronic Automation (DEA), a small robot producer with an outstanding technological capability which has been a source of a number of licences to large US companies but unable to bear the burden of rapidly increasing capital and R&D requirements. STET now also cooperates with IBM in this area.

The large state-owned French electronics groups have entered the field of robot development and production, although their output still is rather low. Auxilec and Sodeteg, part of THOMSON, develop all kinds of flexible manufacturing systems, including robots, in close cooperation with Dainichi Kiko (Japan) and DSR (UK). CGE also entered the field supported by international agreements: the subsidiaries ACB and CGMS, which develop different lines of flexible automation equipment, cooperate with OKK, Sanyo Seiki and Toshiba Seiki from Japan. In 1983, Alsthom–Atlantique, a subsidiary of CGE, acquired CEM, until then a subsidiary of the Swiss company BROWN BOVERI & CO. SCEMI is the robot division of CEM and produces assembly robots. An agreement was signed with Yaskawa according to which Yaskawa Electric helps CEM with the production of large assembly-carrier robots, and SCEMI can sell its small robots in Japan via Yaskawa.[40] MATRA, finally, is engaged in robot production through its division 'contrôle et automatismes' which unites a range of firms engaged in flexible automation. MATRA and RENAULT agreed on a joint marketing strategy leaving the market for small assembly robots to MATRA and reserving the market for robots that can handle heavier pieces to RENAULT, while joining efforts to increase robot exports.[41]

The Swedish firm ASEA is the largest European merchant manufacturer of industrial robots[42] and claims a market share of about 30 per cent of

---

[39] *The Economist* (London), 23 May 1981.

[40] *Japan Economic Journal* (Tokyo), 11 January 1983.

[41] *L'Usine Nouvelle*, 9 June 1983.

[42] In 1985, it had only installed 60 robots in its own factories, which is well below 10 per cent of all robots sold in the open market.

the European market. Instrumental to its diversification strategy towards robotics was the takeover of the robot operations of Electrolux in 1982, and in 1984/5 of the Norwegian firm TRALLFA, which specialized in welding robots. ASEA can be regarded as the only real European 'Triad Power'[43] in industrial robots since it possesses robot production and research facilities in Japan and the United States as well as in Europe.

In short, most major European car manufacturers have opted for the strategy to develop an in-house capability in robotics or cooperate closely with their suppliers. It is interesting to see that the car manufacturers were very early with their robot activities and have developed very advanced versions mainly for captive use (although FIAT and RENAULT have capitalized on their knowhow by selling robots to other firms). Major electronics firms, however, which have a certain robot capability, have entered the scene later and had to rely more on Japanese technology.

2.5. COMPUTER AIDED DESIGN (CAD): KEY TO FURTHER PROGRESS

With the larger processing power of Computer Aided Design (CAD) systems the technology has become a vital input for all the other core technologies. It is mandatory for the further development of very complex products not only in the field of microelectronics (new integrated circuits, robots or computer generations), but also for advances in biotechnology (protein engineering). CAD reduces the costs of design even of more simple end-products and shortens lead times considerably.

Although American companies initially took the lead in CAD product development, the CAD industry in the United Kingdom developed rapidly too, stimulated by early action of the government which established a Computer Aided Design Committee as early as 1967. At the end of the 1960s firms like FERRANTI, RACAL, ICL and MARCONI sold their own 'CAD' hardware. The merging of the UK computer industry into ICL at the end of the 1960s to obtain economies of scale in the competition with IBM ironically resulted in declining growth of the domestic CAD industry.[44] FERRANTI left CAD in the early 1970s and only after 1982 made efforts to enter again. In 1983, GEC's Marconi division took over a domestic CAD system producer, which marks also the return of GEC to this technology.

---

[43] Ohmae (1985).
[44] Arnold (1984), p. 42.

The French industry has also been important in the early set-up of an indigenous CAD industry. The strength of the French industry even resulted in the marketing of French systems by American producers. IBM, for instance, markets a CAD system from DASSAULT, whereas other American firms market AEROSPATIALE's system.[45] With a few exceptions, however, European firms *have never become strong merchant producers of CAD hardware*, certainly not beyond their own home markets. The captive–merchant CAD producers in Europe include DASSAULT, RENAULT, MATRA, ICL, SIEMENS, SELENIA (75 per cent STET group)[46] and also SAAB–SCANIA, VOLVO and ASEA. Most of these firms do not have a very large outlet and use US hardware next to their own products. DASSAULT uses Lockheed–IBM models, RENAULT supplies its own products to, for example, PEUGEOT, but uses also additional US material, SIEMENS uses Computervision hardware besides its own CADIS system, and VOLVO recently has used a mix of own and US turnkey machines for the automation of its production facilities.

The German car industry does not use its own hardware but has developed advanced software. VOLKSWAGEN has its own modelling sytem, DAIMLER–BENZ uses an in-house developed surface modeller in addition to Computervision systems. The firm also purchased a Lockheed simulator (usually used for aircraft) around 1982 to test the look of its designs for new luxury cars. Only PORSCHE seems to have in-house CAD capability without buying systems from external suppliers.[47] ERICSSON is one of the largest CAD users in Sweden, but has only a limited CAD capability.

As a result, the present European market for CAD hardware is dominated by suppliers of US origin.

In all European countries considered in Table 2.3, more than half of the installed equipment comes from US suppliers. Computervision has provided almost every major market with one third of total installations. Since 1981/2 IBM has taken over the leading position of Computervision as a merchant CAD systems supplier.[48]

The *concentration ratio* in hardware supply on the European market thus is very high and differs hardly from the situation in the USA, where the five largest suppliers (IBM (21 per cent), Intergraph, Computervision,

---

[45] Arnold (1984), pp. 27, 28; OTA (1984b).

[46] In the strategy of the US CAD supplier Auto-Trol the joint venture with Selenia in 1984 was an important step for reaching the European market.

[47] Arnold (1984), p. 30.

[48] Other major US suppliers are Applicon, Calma (General Electric) and, to a lesser extent, Gerber, Intergraph and Auto-Trol. Applicon was taken over by Schlumberger which marks the French oil equipment firm's aspirations in this area too.

Table 2.3 Estimated share of suppliers in CAD Stock, 1981/2, in selected European countries (%)

	France (1982)[1]	Germany (1981)[4]	UK (1981)	Sweden (1982)	Norway (1983)
Computervision (US)	36	32	29	31	36
Other US suppliers	26	48	45	34	13
National suppliers	27[2]	9[5]	34[7]	4	37[9]
Other European firms	11[3]	9[6]	—	30[8]	7[10]
Unspecified	1	—	—	—	7

Source: compiled from Arnold (1984).
[1]Excluding IBM.
[2]Mainly MATRA Datavision (6%) and Secmai (12%).
[3]Racal Redac (11%).
[4]Excluding IBM and Calma (General Electric).
[5]Mainly SIEMENS CADIS (85%).
[6]Mainly PRIME/CIS (US/UK) (8%).
[7]Mainly RACAL (15%) and Quest (10%).
[8]Mainly RACAL (21%).
[9]Mainly SI (21%), which stands for the Central Institute for Industrial Research, a state-funded research institute.
[10]i.e. RACAL.

General Electric, Mentor) supplied 62.4 per cent of the total market in 1985. Most of the European firms seem to concentrate on the development of CAD software, mainly for in-house use. Through licensing agreements and direct takeovers, this experience, however, has been tapped by US firms as in the already mentioned cases of MATRA and Aerospatiale. Other examples include the takeover of Shape Data (UK) by Evans & Sutherland (USA) and of Cambridge Interactive Systems, which had a substantial market share in the UK and Germany, and Grado (GER) by Computervision.[49] AKZO Engineering, in cooperation with the UK governmental CAD centre and a private firm (Pipework), developed a very advanced CAD software package for use in process plants and licensed the system to a number of large US users.[50] All major European aircraft producers (next to the French firms, especially FOKKER and MBB and BRITISH AEROSPACE) have a capability in CAD software, since this has become an important asset in their competition strategy. The same can be said of the car manufacturers. The electronics firms investing in VLSI also experienced CAD as an important bottleneck in their development, witnessing their increased investments in this area.

In general, most European multinationals cannot supply turnkey CAD systems. Most of the hardware on the market is provided by US suppliers.

[49] See OTA (1984b), p. 276.
[50] See Kaplinsky (1982), pp. 80, 99.

But whereas these US firms previously consisted of a number of start-up firms, nowadays they consist of large US computer manufacturers supplying hardware *and* software. In the next decade it will become clear whether the concentration of activities of most European multinationals on the software side of CAD has been a viable strategy.

2.6. SOFTWARE: PRIORITY FOR CAPTIVE PRODUCTION

It is not surprising that all major European multinationals have their own software capability. Especially data-processing and telecommunications firms dedicate increasing parts of their total R&D spending to software. The practical integration of systems and sectors (see Chapter One) is predominantly carried out by new generations of software. Large *suppliers* of electronic systems therefore cannot abstain from investing in software to make their systems work. Major US computer hardware firms currently dedicate around 50 per cent of their total R&D spending on software. This will probably increase to 75 per cent by the late 1980s.[51] The latter figure has already been reached for the digital public exchange part of the telecommunications equipment industry, which is estimated to spend around 80 per cent of its development costs on software.[52] Microprocessor manufacturers dedicate more than 50 per cent of development costs to software. This percentage rises steeply with the launch of new microprocessor generations.[53] Of the 30 000 researchers and technicians with SIEMENS, 10 000 are employed in software development. The firm is now trying to start software 'factories' parallel to large Japanese electronics firms to institute more efficient ways of 'producing' software[54] by increasing the division of labour and applying productivity-enhancing technologies such as intelligent workstations, Artificial Intelligence-based debuggers and modularizing the production and storage of debugged code. Software becomes increasingly a 'manufacture'.

For large *users* of electronic equipment (i.e. all European multinationals) it is of strategic importance to have at least a minimum in-house software capability for their own automation strategy. The latter does not mean, however, that European multinationals develop their own software

---

[51] Department of Commerce (1984b), p. 59.
[52] Interview with Ericsson, 1985.
[53] See OECD (1985d), p. 98.
[54] *Financieel Dagblad*, 9 July 1984.

packages exclusively themselves. Probably with the exception of some of the large French and English firms, the supply of packaged software in Europe has been dominated by US firms (IBM having a share of almost 24 per cent in this market segment in 1983).[55] It has been estimated that 80 per cent of packaged software in operating systems on the French market is of American origin. Only in applications which depend far more on national peculiarities, the national content is larger.[56] Even electronics firms like PHILIPS and ERICSSON have chosen the COPICS software package of IBM for their internal logistics automation.[57] PHILIPS first had experimented with an independent software house.[58] Other European firms using IBM *mainframe computers* will also have a strong incentive to use IBM software packages for their manufacturing resource planning, because it complies with their mainframe standard. As a consequence IBM is the second merchant software firm in Europe. But operating systems like COPICS of IBM often lack important modules which have to be developed further by the individual firms, in close cooperation with the provider of the equipment and software.

ICL, SIEMENS and BULL which produce and use their own types of mainframe computers are likely to adopt more quickly non-IBM software standards. ICL, for instance, uses its own production control software to automate its circuit board assembly line, a core activity. These firms, however, together with NIXDORF, PHILIPS and OLIVETTI, do not publish separate data on their software revenues.[59] Of these firms only ICL has established a merchant software subsidiary (ICL Consultancy and Training) offering software services to outside users. The firm is the second supplier in the UK of software services, with a 1982 turnover of $59 million.[60] In February 1985 these more or less 'independent' firms (i.e. ICL, NIXDORF, BULL, OLIVETTI, SIEMENS and PHILIPS) agreed to back a counter-standard to that of IBM in common applications software. It is typical of the rather weak position of the European firms in standardized software packages that the Unix-standard they agreed on was developed by another US company: AT&T![61]

---

[55] The software market is usually analysed on the basis of the mode of delivery of software to users: *packaged software* refers, then, to a standard program which can be used by a wide range of users without major modifications and *custom* software is especially made for the user's needs; another distinction is between systems and applications software.

[56] OECD (1985d), p. 72.

[57] Interviews.

[58] See Wit (1985), p. 56.

[59] Department of Commerce (1984b), p. 36.

[60] Over 20 per cent of ICL 1985 revenue was derived from software. The company claims to carry out more advanced software development than the whole UK software industry. Direct correspondence, December 1985.

[61] *Financial Times*, 19 February 1985.

Other major European firms with relatively important software subsidiaries are: SCICON (BRITISH PETROLEUM), UCSL (subsidiary of UNILEVER, taken over by EDS in 1983), THORN–EMI, THOMSON–CSF and SEMA–MATRA. All have considerable national market shares. It is therefore not surprising that the concentration ratio of French and British software industry is high. In France, the top ten software firms are estimated to receive over 70 per cent of the revenues whereas those in the UK received almost 80 per cent of the revenues in 1982.[62] Most of these firms are tied to their national markets, especially because of language barriers. CAP Gemini Sogeti (France) and SCICON are the only two European firms with a substantial market share in more than two European countries. The other multinational suppliers are subsidiaries of US companies.

It has become more common for the larger firms to use externally supplied packaged applications software (in addition to the systems software which is most of the time sold together with the hardware). Large firms tend to cooperate with the most successful software houses instead of just buying the packages. Thus BULL and SIEMENS (and also IBM, for example) were happy to cooperate with one of the US leaders, Microsoft. This type of cooperation takes the shape of second sourcing agreements as in microprocessors. Research on big software users in Germany underlines this pattern: the share of in-house developed software declined from 74 per cent to 58 per cent in the 1978–85 period. The share of standard programs used by the large firms at the same time increased from 6 per cent to 27 per cent.[63] Much of the latter type of software will be provided by external suppliers since the share of external *custom*-built software hardly changed.

In short, many European firms have complete captive software capability. Most of these are in the non-electronics sector and use their programmers for their in-house automation strategy. They also, however, extensively buy both packaged and custom software from outside suppliers. The software capability of most larger electronics companies seems also to be exclusively of a captive nature in the sense that they have not established independent software marketing initiatives. Instead, the large firms make increasing use of external package software and cooperate with smaller firms. This runs parallel with the pattern in microprocessors

---

[62] For France: *Les Echos*, 21 September 1983; for UK: Quantum Science Maptek Europe, *Seventh Annual Survey of the Computing Service Industry*, 1983, cited in OTA draft, Exhibits XVII and XVIII. For the rest, the European software industry is comprised of a large number of small and medium sized firms producing custom software. Often these software houses started as spin-offs of the automation departments of larger companies. Chris Sprangers and Henk Tolsma, Software in Nederland, *Intermediair*, 11 May 1984.

[63] OECD (1985d), p. 187.

and is an important reason why the *concentration ratio* in the European merchant software industry still is low. Large firms have an interest in keeping the software industry as competitive as possible. Only when very successful corporations in a specific software package appear which set the standard for that market segment (like Microsoft) they either strike a cooperation deal or acquire the firm. As a result the ten largest merchant software houses account for less than 20 per cent of the European market.

2.7. BIOTECHNOLOGY: TARGET AREA FOR PHARMACEUTICAL, (PETRO)CHEMICAL AND FOOD-PROCESSING COMPANIES

The concentration of companies from various backgrounds on the same core technologies is perhaps most obvious in the case of biotechnology. Almost half of the major European multinationals (see Table 2.5, pp. 68 and 69) have become active in this area. Companies of such different branches as the pharmaceutical industry, the chemical industry, the oil industry, the mining and processing industry, the food industry and the engineering industry have tried to get a foothold in this technology, either because they realized that applications of biotechnology could fundamentally affect their traditional product lines, or because they saw a chance to profit from a strong position in one of the neighbouring fields to conquer a share in the newly developing markets as well. Companies either needed up-to-date knowledge in biotechnology as an input for improvement of existing products and production processes, or wanted to exploit their process knowhow or marketing strength in new fields. There have hardly been any 'biotechnology companies' in the past, with the exception of GIST–BROCADES in the Netherlands, which has been an important producer of yeast, and the Danish company NOVO, the world's largest producer of enzymes. Companies actually active in biotechnology come from different fields. The large companies are the main carriers of the development of biotechnology in Europe, not small start-up companies as in the United States.

The closest to modern biotechnology has been the *pharmaceutical industry* given its intrinsic interest in the life sciences. But though important activities of European pharmaceutical companies do exist in the field[64], these have been more cautious than American companies. Three reasons may be responsible for this:

[64] European Federation of Pharmaceutical Industries' Associations (1984).

(a) The development of biotechnology in the United States was massively financed by the National Institute of Health (which has been for biotechnology what the Pentagon was for microelectronics). This implied that medical applications were dominant in biotechnology research. There was no similar concentration of efforts in Europe.

(b) American pharmaceutical companies have been more strongly oriented towards antibiotics[65] than many of the large European pharmaceutical companies. In Germany, for example, there is still a much lower consumption of antibiotics per capita than in the United States, and a much lower share of antibiotics in the total consumption of pharmaceuticals.

(c) The largest European pharmaceutical companies are at the same time the largest chemical companies, which is not the case in the United States. Some of these companies had gained rather negative experiences with the development of Single-Cell Protein (SCP) during the 1970s and thereby became disillusioned with regard to the potential of biotechnology for a while.

Because the largest pharmaceutical companies and the largest chemical companies are almost identical, it would not make sense to describe these two groups separately. Several of these companies have become involved in biotechnology more as chemical than as pharmaceutical companies. BAYER is the world's largest producer of pesticides. In order to remain at the forefront in this field, the company cooperates with the state financed Max-Planck-Institut für Züchtungsforschung and with the Genetics Institute of the University of Cologne in biotechnology. Expenditures on biotechnology increased quickly: from DM150 million in 1982 to DM180–90 million in 1983.[66] Besides, BAYER had taken over the American company MILES in 1977, which is the world's third largest enzyme producer (after NOVO and GIST–BROCADES) and is majority owner of the American biotechnology company Molecular Diagnostics.

HOECHST entered biotechnology research after the SCP fiasco (see Chapter Five, p. 173) through a large research contract with the Massachusetts General Hospital of Harvard University in Boston. The company is determined to play as large a role in biotechnology as it does in the chemical and pharmaceutical industry in general. The company considerably expanded its own laboratories for genetic research at Frankfurt and spent about DM90 million on biotechnology research, for example, in 1982.[67]

---

[65] The 'second generation' biotechnology. See Sharp (1985b), p. 15.

[66] Sehrman (1983) *Biotechnik: Der Weiche Riese*, Die Zeit, 11 March.

[67] About 25 per cent of this amount was spent on basic research, 40 per cent on product development and 35 per cent on process development. *Handelsblatt*, 4 July 1983.

Of the three German chemical giants, BASF (which is the least active in pharmaceuticals) is obviously the most cautious in the field. It invested about DM50 million in a new biotechnology laboratory and cooperates with the University of Heidelberg in genetic research.[68] The large Swiss chemical and pharmaceutical companies CIBA–GEIGY and HOFFMANN–LA ROCHE were very fast in acquiring a research capacity in biotechnology—to a large extent (as in the case of German companies) in the United States. CIBA–GEIGY opened a biotechnology laboratory at Basle for pharmaceutical applications and another one in North Carolina (USA) for applications in agriculture. HOFFMANN–LA ROCHE spent nearly as much on biotechnology research in 1981 as each of the two largest spenders among American pharmaceutical companies, SCHERING–PLOUGH and ELI LILLY.[69]

In Britian, ICI proved to have a strong position in fermentation technology, when it installed the largest bioreactor in the world (for the production of single-cell protein).[70]

The Italian concern FERUZZI in no time became one of the most important biotechnology corporations in the world. It has become the second largest private corporation in Italy (after FIAT and before OLIVETTI), and is the third largest food concern in the world (behind NESTLÉ and UNILEVER). FERUZZI, however, concentrates its activities much more on biotechnology-related areas (e.g. ethanol and starch production). It also bought about 40 per cent of the shares of the chemical concern MONTEDISON, which has given priority to biotechnology (especially in applications in pharmaceuticals) since a major turnaround of the company which brought it back to profitability in 1985.

The chemical industry used to make much more use of fermentation products at the end of the nineteenth and early in the twentieth century, but the fermentation route to a large number of products became uneconomic when petrochemical processes were developed. The actual developments of genetic engineering, however, could render fermentation processes far more efficient and may reverse this trend in many cases.[71] There are five reasons why *petrochemical companies* are interested in biotechnology:

(a) Part of their production may be replaced in the future by products that result from application of biotechnology. To compensate for these

---

[68] *Industriemagazin*, **3** (83), 47.
[69] OTA (1984a), p. 74/5.
[70] Its 'Pruteen' plant was commissioned in early 1980. Cf. Scott (1984), p. 235 f. Pruteen did not become a market success, and ICI has been looking for quite a while for a buyer of the production technology.
[71] See Cohendet (1984).

losses, the companies are interested in using their marketing channels to produce and market the substitutes themselves in order to stay in business.

(b) The large oil companies are no longer oil companies only. They have become *energy companies* with large stakes in nuclear energy and coal, besides oil—but also with considerable interest in all forms of 'alternative energy', including the use of biomass and its conversion to biogas or fuel alcohol.

(c) Even for the core of their present business—oil—biotechnology can be relevant. About two thirds of world reserves of light oil can only be produced with 'tertiary methods of recovery'.[72] Biotechnology can play an important role in this respect. It can help to produce those polymers that can change the underground structure of oil reserves in a way that enables more oil to be pumped to the wells.

(d) Biotechnology can also help to develop new markets for oil: oil and gas were used as possible raw materials for SCP production.

(e) Finally, the largest oil companies (especially SHELL) have also become large agrochemical companies with the same interest in biotechnology as other chemical companies, as described above.

The largest European oil companies all have become very active in biotechnology. BRITISH PETROLEUM very early invested large sums in SCP production but had to write off this investment and biotechnology became anathema for the company for quite a while.

SHELL experimented with biotechnologically produced polymers for tertiary recovery. Its main interest in biotechnology, however, is as an agrochemical company.[73]

The fourth largest company, ELF–AQUITAINE, perhaps is the most interesting. It is the largest French biotechnology company. Its biotechnology subsidiary SANOFI is used as a spearhead to promote biotechnology in France. Within a few years, the company bought many smaller firms in the field and restructured the industry: SANOFI, of which ELF–AQUITAINE owns about 60 per cent (with the rest held by the public), was set up in 1973 for developments in the health and personal care sector. From 1981 on, it became the main pillar of ELF's efforts in 'life chemicals', grouping most of ELF's shares in

---

[72] Traditional primary and secondary oil recovery leaves about two thirds of the original oil in place in the ground. When the natural pressure (primary recovery) is no longer sufficient to bring the oil to the surface, and the injection of water or gas (secondary recovery) is no longer enough to redress underground pressure and drive the oil to the well, chemicals have to be added (or eventually produced by bacteria underground) that change the oil-bearing strata in such a way that more oil can be produced. See Moses (1983).

[73] See Ruivenkamp (1986).

pharmaceuticals, animal products and plant chemicals. ELF and SANOFI have become very active in acquiring the knowhow they missed by buying up smaller companies (mostly in France) or by participating in other companies (mostly abroad). In this way, ELF–AQUITAINE probably has become one of the most diversified biotechnology companies in Europe, with activities ranging from seeds, agrochemicals, animal products, dairy technology to aromatics, pharmaceuticals and biomass energy.[74]

Besides pharmaceutical, chemical and oil companies, *food producers* have become very interested in biotechnology. The reasons for their interest are obvious. Biotechnology helps food companies to use a broad range of alternative raw materials for the same end-products and thus makes them more independent from specific suppliers. Biotechnology provides companies with many technologies to produce new fragrances and flavours, new stabilizers, textures and preservatives. Biotechnological manipulations of ingredients help to prolong shelf life of many food products without adding preservatives. This gives large companies with higher economies of scale a competitive edge with regard to smaller, regional companies that hitherto have been able to survive because of the need for fresh products.

NESTLÉ and UNILEVER, the two largest European food companies, both have become very active in biotechnology. While little is known of the concrete biotechnology research activities of NESTLÉ, UNILEVER has become famous for its successes in *tissue culture*, especially of oil palms, which will provide the company with cheaper edible oils as raw materials. UNILEVER underlined its strong interest in biotechnology by its effort to take over NAARDEN INTERNATIONAL in 1986, a smaller Dutch multinational specializing in the production of fragrances and flavours.

Europe did not experience a similar development to that of Japan, however, where food companies, on the basis of their considerable knowhow in fermentation technology, entered the field of pharmaceuticals with fermentation products.

Besides the two large food producers UNILEVER and NESTLÉ, many others have become active in biotechnology, especially the wine producers and breweries (like MOËT–HENNESSY in France, HEINEKEN in the Netherlands and Denmark's UNITED BREWERIES) and sugar refineries (as the DANSKE SUKKERFABRIKKER or the Dutch CSM). These companies either cannot be regarded as multinationals, however, or they lack a research orientation and therefore cannot be regarded as 'high tech' companies. For these reasons, they will not be included in our analysis.

[74] Wiltgen (1987), pp. 23–5.

Finally, some engineering companies have also acquired a stake in biotechnology, because they produce the equipment for large-scale fermentation processes. Among these companies are ALFA–LAVAL in Sweden, which has a strong world market position in equipment for milk processing, and KRUPP.

Other multinationals got a stake in biotechnology not so much because of a specific link with their traditional lines of business, but as a move of diversification into a promising field of the future. A case in point is VOLVO which in 1985 became the single largest shareholder of PHARMACIA, the largest Swedish biotechnology company. VOLVO described this acquisition as part of a long-term move into the sector.[75] Since there is no link with VOLVO's main line of activities, this move finally may end up as the takeover of TRIUMPH–ADLER by VOLKSWAGEN.

2.8. NEW MATERIALS

'New materials' represent such a broad category that it is not surprising that almost *all* major European multinationals have some competence in this area. Many firms see new materials as a strategic new input in their end-products or their production processes.[76] So *automotive producers* have become engaged in research on ceramics to be used in new engines and in specialty steels and synthetic materials as a substitute for steel bodies. With regard to the *electronics firms*, research on new substitutes for silicium (gallium arsenide, for example), the basic material for integrated circuits, has priority. Progress in the use of other synthetic materials is also of vital importance for the wider area of 'optronics'.

Increased competition in the *(petro)chemical* and *metal* industries from the Middle East and from South-East Asia at the lower end of the market has spurred their restructuring away from bulk and commodity products towards areas with a higher value added, such as new materials. The chemical companies are aiming at specific synthetic materials (engineering

---

[75] Its holdings already included (a) a 34 per cent stake in Senessons which in turn controls the Leo–Ferrosan pharmaceutical company and the Gambro medical equipment group, (b) a 22 per cent share in the biotechnology and sugar company Cardo, and (c) at least 15 per cent in KabiGen, one of the world's leading producers of human growth hormone. *Financial Times*, 5 February 1985. Volvo sold 85 per cent of the shares of AB Leo for SEK3 billion to Farmacia in July 1986 and in turn increased its holding of the latter from 26.8 per cent to 31.0 per cent.

[76] Cf. Roobeek (1987a).

plastics, synthetic fibres) and composites. Thus AKZO invested more than DFL500 million in a new synthetic fibre (aramide). Due to the fact that the Dutch Government subsidized around half of the investments, AKZO could develop the new fibre without seeking for partners.[77] DSM is a good example of a company which has changed its corporate strategy considerably since the beginning of the 1980s in order to move into higher value added areas. In 1984 it centralized all high tech research projects in one department (with more than 1000 researchers). At least five out of the eight projects the company aims at deal with new materials: polyetheen fibre (comparable with AKZO's aramide fibre), gel-technology to produce the fibre, high-performance composites, high-performance synthetic materials and ceramics.[78] For firms such as DSM, which traditionally were almost exclusively active in bulk chemical production with a low R&D intensity, this diversification will take considerable time.

Further advanced on this way is ICI, a much larger traditional bulk chemical producer. 'Committed to being an innovative company', ICI has developed a number of new materials, ranging from a new sort of cement with low-volume porosity which is approaching the flexural and tensile strength of aluminium, wood and some plastics, to new thermoplastics with good heat-resistant properties, which can be used for high-temperature aerospace applications, to strong and safe inorganic fibre as a replacement for asbestos.[79]

In contrast to ICI, HOECHST had never specialized in bulk chemicals. Nevertheless, the company makes great efforts to develop more technical chemicals. HOECHST raised its investment portfolio from DM1.9 billion in 1984 to DM3 billion in 1986. These investments explicitly aim at what the firm calls 'High Chem' activities which correspond with the high value added, difficult-to-copy products with attractive profit margins. Targeted areas in the company's strategy will be pharmaceutical products and processes, foils, but increasingly also fibres and organic chemicals. A related acquisition in this area concerned Ceramtech, a firm engaged in research and production of technical ceramics.[80] A cooperation deal has also been struck with the US 3M corporation for the production of fluor elastomers (high-temperature resistant rubber).

The steel industry has been slow in comparison to the chemical industry in changing its strategy towards new materials. This can be explained by the fact that the chemical industry has been more innovative, much less

---

[77] Since 1985, AKZO has been fighting a fierce patent battle with the American producer, Dupont, of a comparable fibre which has important consequences for the successful marketing of the AKZO fibre.

[78] *NRC-Handelsblad* (Rotterdam), 24 July 1985.

[79] ICI Europe (no date).

[80] *De Volkskrant* (Amsterdam), 20 November 1985.

mature and producing more 'high tech' goods than the 'old' steel industry. The role of the chemical industry in the development of new materials, also as direct substitutes for steel, can thus be expected to be larger than that of the steel giants. More innovative steel producers like KRUPP in Germany or HOOGOVENS in the Netherlands, however, are investing in areas close to their bulk business, not only in specialty steels but also in composites (high strength, low alloy).

In sum, a certain pattern of *specialization* develops in Europe. The car manufacturers see the new materials they develop primarily as strategic inputs for their end-products and as a means for further automation of production. They have targeted ceramics first of all. The chemical and steel industries try to diversify into 'new' materials in order to put more emphasis on the higher value added end of the market. The chemical companies are especially active in the areas of composites and synthetic materials such as new plastics. In a number of areas the strategies overlap and a growing network of cooperation between, for example, chemical and automotive firms can be witnessed.

2.9. THE INCREASING IMPORTANCE OF R&D FOR CORE TECHNOLOGIES

The flexibility of changing the product range of firms in the direction of the core technologies basically must come from the R&D departments. One would expect that the new generation of core technologies necessitates a bigger commitment to R&D[81] than previously. The analysis of three indicators of the higher priority attached to knowhow (i.e. R&D) will show in how far this is the case.

First, we *compare the R&D budgets and the capital expenditures* of a sample of European multinationals (see Table 2.4 for figures on R&D in the 1975–85 period; for capital intensity see Table 3.5). In the 1970s priority was given to investment in fixed capital. Since the beginning of the 1980s, however, a clear trend towards a higher priority for investments in R&D can be observed: in more than half of the firms of our sample, the ratio R&D expenditures:capital expenditures increased in favour of R&D. Most

---

[81] The data on R&D expenditures of European companies generally relate to personnel and laboratory equipment because of their tax deductibility. This somewhat underestimates expenditures on R&D, but since the distortion is similar for different companies, a rough comparison between companies is possible. The figures on R&D have to be used with care, since the same tax deductibility might induce firms to inflate their 'official' R&D figures!

Table 2.4 R&D expenditures of major European multinationals, in ECUs (millions) and % of sales

	1975		1980		1984		1985	
	ECU millions	%	ECU millions	%	ECU millions	%	ECU millions	%
SHELL	200.0	0.6	369.6	0.6	664.4	0.6	804.8	0.7
BP	44.6	0.3	103.7	0.3	394.5	0.6	496.3	0.6
ELF	71.4	1.0	127.7	1.0	334.7	1.3	n.a.	n.a.
RHÔNE-POULENC	181.9	4.5	232.9	4.5	359.0	4.8	397.5	3.7
AKZO	99.5	4.0	149.3	3.3	232.6	3.5	263.3	0.8
DSM	n.a.	n.a.	31.4	1.3	69.3	0.8	79.6	3.2
ICI	198.2	3.6	354.2	3.7	506.8	3.0	648.7	3.0
BASF	207.9	3.5	393.4	3.6	554.9	3.1	644.1	4.6
BAYER	263.4	4.5	492.5	4.3	874.0	4.5	957.8	5.4
AGFA-GEVAERT	n.a.	n.a.	103.3	5.9	168.0	5.4	182.0	4.9
HOECHST	305.0	4.5	515.4	4.3	812.3	4.4	934.9	1.6
DEGUSSA	24.3	2.0	49.1	1.4	73.7	1.5	85.3	9.2
CIBA-GEIGY	224.8	8.0	402.5	7.9	778.1	8.3	901.4	4.9
GIST-BROCADES	11.5	4.2	21.7	4.4	29.7	4.2	35.8	1.5
UNILEVER	134.9	1.1	264.8	1.4	396.3	1.5	403.4	6.7
PHILIPS	638.0	7.4	998.2	6.9	1427.9	6.7	1597.4	8.8
SIEMENS	528.0	8.5	1188.5	9.0	1697.7	8.3	2154.4	7.9
AEG	247.9	5.9	394.6	6.9	353.4	7.2	382.4	4.8
BOSCH	112.1	4.7	240.5	6.7	404.8	6.4	456.5	3.7
ASEA	n.a.	n.a.	110.5	5.2	199.7	3.6	230.5	

ERICSSON	99.6	6.4	172.2	8.3	361.7	429.9	8.0	8.6
BULL	n.a.	n.a.	106.5	9.9	205.1	235.5	10.4	9.9
THOMSON	348.4	n.a.	664.5	10.7	858.7	n.a.	10.3	n.a.
CGE	n.a.	n.a.	277.7	3.6	625.7	n.a.	5.8	n.a.
ICL	n.a.	n.a.	n.a.	n.a.	189.8	n.a.	10.0	n.a.
GEC	n.a.	n.a.	n.a.	n.a.	1049.1	n.a.	11.1	n.a.
BRITISH TELECOM	n.a.	n.a.	159.8	3.8	308.4	n.a.	2.4	n.a.
STET	10.1	3.1	56.9	2.3	326.5	193.9	3.5	4.6
OLIVETTI	14.7	7.3	47.6	7.7	113.5	172.8	3.4	9.8
NIXDORF	n.a.	n.a.	n.a.	n.a.	142.2	616.3	9.9	4.7
FIAT	213.2	3.4	495.2	3.7	484.3	763.0	2.8	3.2
VW	223.0	3.2	436.5	3.5	625.5	763.0	3.1	3.2
DAIMLER–BENZ	42.6	4.0	138.7	4.3	670.2	n.a.	3.4	n.a.
BMW	89.5	1.1?	291.7	3.4	n.a.	n.a.	n.a.	n.a.
RENAULT	n.a.	n.a.	251.1	2.1	473.1	586.5	2.8	4.4
PSA	n.a.	n.a.	270.9	6.7	237.7	n.a.	1.8	n.a.
VOLVO	52.6	1.3	85.0	1.4	507.4	n.a.	3.8	n.a.
SAINT-GOBAIN	27.9	0.9	59.5	1.1	136.8	n.a.	1.5	n.a.
KRUPP	31.2	0.7	n.a.	n.a.	98.3	106.4	1.2	1.3
MANNESMANN	n.a.	n.a.	n.a.	n.a.	n.a.	143.6	n.a.	1.8

electronics companies (SIEMENS, AEG, ERICSSON, THOMSON, CGE, GEC and ASEA) have already had higher expenditures on R&D than on fixed assets, whereas the diversification strategy of some other companies (notably BAYER and VOLVO) have also resulted in an increase of their absolute R&D outlays beyond capital expenditures.

A second indicator is the *R&D:profit ratio*. This ratio has experienced a considerable increase in all firms. In fact, the R&D budget has progressed almost totally independently of any development in (pretax) profits. In the 1980–2 period, when most firms experienced a severe decline in their profits, the R&D expenditures of the firms in our sample increased without exception. The strategic nature of R&D expenditures thus is not only a theoretical credo, but has been practised by *all* large companies in spite of the squeezing effect of lower profit margins. In fact, profit squeezes have urged companies to *raise* their R&D expenditures in order to find an innovative way out of the slump and restore the companies competitiveness. Of the firms in our sample, only THOMSON and AEG in the 1982–3 period slightly lowered their R&D expenditures, having experienced considerable losses in the preceding year(s). For AEG the reason for this decline becomes clear if one considers that it was forced to sell some of its most lucrative (and R&D intensive) business in this period in order to restructure its loss-making operations.

Increases in the *R&D:sales ratio* is a third indicator generally used to indicate shifts in the knowhow orientation of firms. Looking at Table 2.4 we see that the picture with regard to this indicator becomes somewhat mixed, since no general increase is apparent. A bigger commitment to R&D than previously (expressed as percentage of total sales), however, is only necessary for firms which were not yet engaged in what have now become the core technologies, or which operate in market niches. Thus firms already in an advanced position, especially in microelectronics and to a lesser extent pharmaceuticals, have not seen major changes in their already high R&D share.

When we consider the *absolute R&D expenditures* of the sample of European multinationals, three broad categories can be discerned (in 1984/5):

(a) The *super league* of firms which spend more than ECU 1 billion per year on R&D. In 1985 this group consisted only of the highly diversified electronics companies: GEC, PHILIPS and, on top, SIEMENS. This resembles the US and Japanese situation where in each case three electronics firms (respectively IBM, AT&T and UTC; HITACHI, MATSUSHITA and TOSHIBA) belong to the super league of the five largest R&D spenders. Cash-rich SIEMENS has been rapidly increasing its commitment to R&D in the 1980s, aiming at the Factory-of-the-future as well as the Office-of-the-future. It quickly surpassed PHILIPS, which in 1975 was the European front-runner in R&D expenditures. In comparison with SIEMENS, the strategy of PHILIPS is more aimed at only the Office-of-the-future (together with its Household-of-the-future strategy).

(b) The *big league* companies spend between ECU500 million and ECU1 billion on R&D. Most diversified (petro)chemical companies (BAYER, HOECHST, SHELL, BASF, ICI, CIBA–GEIGY) belong to this group, together with two (compared to the 'super league') less diversified electronics companies: THOMSON and CGE. At the lower end of this league we also find five of the automobile producers: DAIMLER–BENZ (before acquisitions), VOLKSWAGEN, VOLVO, FIAT and RENAULT all spent between ECU500 million and ECU750 million on R&D in 1985.

(c) The *little league* firms have generally specialized in specific market segments, like some of the 'smaller' computer firms (BULL, ICL, OLIVETTI, NIXDORF), the telecommunications firms (BRITISH TELECOM, ERICSSON and the STET group, though these have higher budgets than the specialized computer firms), some car manufacturers (PEUGEOT (PSA) and BMW) and many less diversified petrochemical and food companies. At the lower end of this group, firms predominantly engaged in bulk production (KRUPP, DSM) can be mentioned.

The relative importance of R&D for the 'super league' firms has not changed radically during the 1975–85 period. The same can be said for firms in the other 'leagues' not willing to alter their degree of specialization/diversification. The group of firms which have *changed* their business orientation in order to get rid of the status of a specialized (or bulk) producer to become more diversified shows the largest increases in R&D spending (relative to total sales). This group includes firms such as ELF, BP, OLIVETTI, CGE, NIXDORF, FIAT, DAIMLER–BENZ, HOECHST and VOLVO. The rise of VOLVO's R&D spending and rapid diversification in 1983, for instance, brought it from the 'little' into the 'big' league (as opposed to, for example, PEUGEOT which did not seek such diversification and consequently stayed in the 'little league'). The strategies of HOECHST and DAIMLER–BENZ also deserve separate mention since the two companies managed to raise their R&D budgets (even as part of sales) considerably after 1984. The three big takeovers of DAIMLER (p. 29) raised the R&D share in total sales from 3.4 per cent in 1984 to around 5.5 per cent in 1986. The higher priority for R&D at HOECHST resulted in a rise of the share of R&D from 4.4 per cent in 1984 to 4.9 per cent in 1985 and a similar increase in 1986. As a result, both companies which aim at moving into the core of the core technologies, entered the European 'super league'.

Table 2.5 Activities of European multinationals in core technologies

Firm	Chips[4]	Comp.	Robots	CAD	Telec.	Softw.	Biotech	N.mat
SHELL	—	—	X	—	—	X	XX	XX
BP	—	—	—	X	2	XX	X	BC
IRI/STET[6]	XX°	—	XX°	XX[1]	XX°	X	—	C[1]
UNILEVER	—	—	—	—	—	—	BD	ABC
ELF-AQUI.	—	—	X[1]	—	—	X	XX	BC
PHILIPS	XX°	XX	X	X	XX[1]	XX	—	AC
SIEMENS	XX°	XX[3]	XX[1]	X	XX	XX	—	C[1]
VOLKSWAGEN	—	X	XX*	X	—	X	—	AX
DAIML./AEG	X	X	—	X	X	X°	—	AX
BAYER	—	—	—	—	—	X	BAC	ABC
HOECHST	—	—	—	—	—	X	ADC	ABC
RENAULT	—	—	XX	X	—	X	—	ABCD
FIAT	—	—	XX[1]	XX	XX	X	—	A
NESTLÉ	—	—	—	—	—	—	X	—
BASF	—	X[1]	—	—	—	—	XX	BCE
VOLVO	—	—	X	X	—	X	AB	ABD
ICI	—	—	—	X	—	—	ABCDE	ABCE
PEUGEOT	2	—	—	X	—	X	—	A
CGE	2	XX[3]	XX°	—	XX	X	—	AC[3]
SAINT-G.	—	—	—	—	—	—	—	C[1]
CIBA-G.	—	—	—	—	—	X	XX	XX[1]
DSM	—	—	X[3]°	X	—	X	XX	XX
GEC	XX[3]	X	X	XX	XX	X	—	ABC
KRUPP	—	X	X°	—	—	X	XB	ADE
THOMSON	XX°	X	—	X	2	X	—	2[5]
RHÔNE-P.	XX	—	—	X	—	—	XX	XX
BOSCH	X*	X	X	X	X[1]	X	—	AC

MANNESMANN	—	X	—	X	X[1]	X	—	XX
AKZO	XX†	—	—	X	—	X	XX	BCE
THORN–EMI	X*	—	X[1]°	X	XX[1]	XX†	—	BC
ASEA	X*	X*	XX†	X	—	X	—	AB
ERICSSON	X*	X°	X[1]	X	X	X	—	C
OLIVETTI	X	XX[3]	—	X	X[1]	X	—	—
BULL	X°	XX°	X	X	X[1]	XX	—	C
ICL/STC		XX[1]	—	XX	XX	—	—	XX
DEGUSSA	—	—	—	—	—	—	XX	—
GIST–BR.	—	—	—	—	—	—	XX	—
BRITISH T.	XX[1]	XX	—	X	XX	X	—	X
MATRA	—	—	XX°	XX	X	X	—	—
NIXDORF	XX	XX	—	X	XX	XX	—	—
MONTEDISON	—	—	—	—	—	—	XX	BC

XX Major activity
X Competence
* Predominantly captive production (above 60%)
° Captive–merchant production
† Predominantly merchant production (below 15%)

Key for biotechnology:
A (R)DNA
B Fermentation products
C Cell fusion
D Enzymes
E Others

Key for new materials:
A Fine ceramics
B Composites
C Synthetic materials (plastics, optical fibres)
D Specialty steels
E Others

[1](Partly) in joint venture or in cooperation with other firms
[2]Withdrawn in 1980–5 period
[3](Partly) under licence
[4]If we specify the in-house use of chips, the following estimates can be done for major firms: SIEMENS 22% of all components (annual report 1984), PHILIPS 15% (interview), THOMSON 15%, SCS–ATES (STET) low percentage own use. HAFO (ASEA) and RIFA (ERICSSON) provide 20% of the local components market, but in integrated circuits their production is mainly captive.
[5]Cable operations of Thomson were transferred to CGE, but probably new materials for integrated circuits are left with the firm.
[6]Other IRI subsidiaries, however, are involved in computers (IRI–Finmeccanica), software (IRI–Finsiel) or biotechnology (IRI–Finmeccanica).

For many firms the reorientation towards core technologies and higher research intensity, however, is not reflected adequately in higher R&D expenditures, since other steps have been taken besides in-house R&D to get a stronger position in the new technologies. Companies have strengthened their engagement (a) by the increased use of technology transfer through second sourcing, (b) by minority shares in small high tech companies, (c) by the more intensive use of cooperation agreements (technology exchange, joint ventures, original equipment manufacturing, licences) in general (see Chapter Seven) which spreads the cost of research and development over more partners and (d) by increases in government support (see Chapter Six). Without these arrangements, companies would have had to increase their R&D expenditures even more.

2.10. A DIVISION OF LABOUR BETWEEN EUROPEAN MULTINATIONALS

In the introduction to this chapter the strategic choices which the large firms face with regard to their product structure and technologies where listed. This chapter has provided material on the strategies chosen by European multinationals as an answer to four of the six questions they have to address. Two questions relate to the consequences of the use and production of core technologies on the *organization* of the firm. These will be dealt with in the next chapter.

Table 2.5 summarizes the main decisions taken by 41 multinationals[82] with regard to the targeted core technologies.[83] The general impression from this overview is that, in any area, almost half of the firms chosen

---

[82] Most firms whose choices have been described in this chapter are included in the sample of 41 European multinationals listed in Table 2.5, which serves as a reference group in this book. In principle, the multinationals with a European headquarters among the first 100 firms in the well-known Fortune ranking were chosen. Twelve European firms among the first 100 were left out because of a comparatively low international presence, a low profile in core technologies or the predominantly trading nature of their business. These firms are: ENI, Total Cie, Française des Pétroles, BAT Industries, Thyssen, Petrofina, Ruhrkohle, National Coal Board, Michelin, Preussag, British Leyland, Grand Metropolitan and Dalgetty.

Other companies have been added not for their absolute size but because of the importance of their activities in high technology sectors as already analysed in this chapter. So firms like Nixdorf (number 408 on the *Fortune* list) or the Dutch Gist–Brocades (not among the first 500 firms) are included. Some other companies might have been suitable candidates as well, but pragmatic reasons obliged us to keep the size of the sample down. Since it is not meant to be a representative sample in any strict statistical sense, but serves mostly illustrative purposes, we believe this pragmatism can be justified. Where appropriate in the remainder of this book, multinationals active in core technologies but not included in our basic sample will be included as well.

[83] Table 2.5 is based on a questionnaire to which 27 of these firms replied.

have some competence. This means, for instance, that oil companies like SHELL and ELF surprisingly have competence in robotics (or 'mechatronics' as SHELL calls it) not as the result of a diversification strategy but because of the use of submarine robots in the off-shore business. There is a clear dividing line between electronics companies busy in semiconductors, computers and robotics on the one hand and (petro)chemical companies entering biotechnology on the other. There seem to be no large European companies yet that tackle the field where the two clusters of core technologies overlap: bio-informatics. A certain 'link' is formed by new materials, an area in which almost all firms have a vital interest. If we consider specific areas of new materials (such as ceramics), some kind of division of labour can also be found (see section 2.8). The interest in CAD and software in three quarters of the companies indicates that these areas might be considered as essential for all companies. A minimum in-house competence is vital to stand up against international competition. Hardly any European firms had these areas as *major* (merchant) activity, but all developed capacities for in-house use. Firms like BAYER, which have no major activities in the microelectronics cluster, still already spend 20 per cent of their R&D budget on information technology.[84] These expenditures are particularly aimed at areas such as software and CAD which relate more to the process innovations referred to in the next chapter.

Depending on their position in the technology cluster, firms opt for diversification or specialization. This has induced many diversified companies to sell off part of their so called *non-strategic* business in order to become or stay a major competitor in core technologies. An example is THORN–EMI which divested its cinema business and metal industries division in the 1984–6 period in order to reserve funds for more strategic activities. Most vertically integrated companies are doing the same with respect to their 'non-traditional' business. At the other extreme there are the oil multinationals, with an enormous cash-flow but with a traditional business reaching the boundaries of its growth potential. A number of these firms have also tried to invest in microelectronics and set up their own data-processing companies. Firms like BRITISH PETROLEUM (and in the United States EXXON and ATLANTIC RICHFIELD) are examples. In general, however, such diversification into the information technology cluster has not been very successful for most of these firms. Often the company 'culture' has not been flexible enough and the newly started activity could not be sufficiently integrated with the rest of the firm's activities. After some time this prompted the firms to withdraw again from these areas.

[84] Bayer, *Annual Report, 1984*, p. 12.

Table 2.6 Picking up the pieces

Specializing firm	Restructured part	Taken over by other specializing firm
Atlantic Richfield (oil)	Annaconda (telecom)	ERICSSON (1985) (previously joint v.)
EXXON (oil)	data processing	OLIVETTI (1984) (European part)
BRITISH PETROLEUM (oil)	share in Mercury (telecom)	Cable and Wireless (1984)
United Technologies	Mostek (semiconductors)	THOMSON (1985)
UNILEVER	data processing	EDS (GM) (1984)
CGE	consumer electr. and military	THOMSON (1984)
AEG	Telefunken (cons. electr.)	THOMSON (1983/4)
AEG	T&N (telecom)	BOSCH, MANNESMANN
THOMSON	telecommunications	CGE (1984/6)
THOMSON	Socapex ('connecteurs')	ALLIED–BENDIX (1985)
SAAB–SCANIA	Datasaab	ERICSSON (1980/1)
Electrolux	FACIT	ERICSSON (1982)
VOLKSWAGEN	Triumph–Adler	OLIVETTI (1986)
PHILIPS	camera division	BOSCH (1985/6)
ITT	telecommunications subsidiaries (SEL, Bell-Belgium, etc.)	CGE consortium (1986)
MATRA	clockmaking division MATRA HORLOGERIE	HATTORI group (Seiko watches) (1986)
THORN–EMI	metal industries division division & cinema business	various companies (1985/6)
PHILIPS	40% share in Polygram (records)	SIEMENS (90%) (1986)
AKZO	American Enka (synthetic fibre)	BASF (1985)
BROWN–BOVERI CO.	CEM (robotics)	CGE (1983)
UNILEVER	Stauffer (chemicals)	ICI (1987)
ICI	Stauffer Chemical's basic chemicals division	RHÔNE–POULENC (1987)

Most firms, instead, have increasingly become aware of the necessity of adopting a diversification strategy in product areas more linked with their own expertise. For the oil companies this is thus more in new materials and biotechnology.

The strategies of different firms come together where firms which invest in core technologies pick up the pieces of other firms (such as the oil multinationals) specializing in other areas. Some examples are listed in Table 2.6. The overall result is a continued concentration of specific core technologies with a few large and relatively coherent firms. On the other hand, the non-strategic activities of some multinationals are taken over by other multinationals for which they have no strategic value in a technological sense either, but where they make them realize higher economies of scale. The interfirm restructuring thus also results in a further *concentration of traditional and 'low tech' activities in the hands of a few firms*, which can better rationalize them and make them more profitable due to more efficient production for larger markets. It gives these firms also the possibility to upgrade these product areas by investing more in R&D to develop higher value added products with higher profit margins. The reshuffling at the lower end of the market also results in a specific division of labour between the multinationals which we will not analyse further in this book.

In the next chapter the consequences of the changed product strategies in the direction of the core technologies for the *internal organization of the multinational* will be considered.

Changing Structure of Multinationals

Multinational corporations become more and more *design, management, assembly, trading* and *service firms,* with large parts of production put out to smaller companies (subcontractors) under their control. They then become more similar to Japanese companies. Multinationals thus exercise considerable influence on the structure of production of national economies which reaches far beyond their legal boundaries.

Developments in new technologies have led to challenges and chances for multinational companies which urge them to reorganize their internal structures as well as their relationship with subcontractors for a number of reasons:

—The integration of technologies (for instance, of data processing and telecommunications, see Chapter One) blurs the boundaries between traditional divisions.

—Efforts to link automation of design, production and administration make a stronger centralization of specific functions necessary.

—The increased research intensity of high value added production implies that R&D departments are given a greater weight and get reorganized themselves.

—New specialization and diversification as well as efforts to reduce inventories and to avoid workers' influence on management decisions have changed the attitude of large firms towards their subcontractors.

The introduction of computer assisted production technologies into the process industries started in the 1970s. In discrete production, however, only in the beginning of the 1980s were large-scale rationalization investments undertaken and a new era in the mode of production took shape. Following the analysis of Coombs, three elements of the (discrete) production process can be mechanized: the manipulation and manufacturing of goods and materials, the transport of products between the manufacturing centres and the control over both activities. Since the industrial revolution in the mid nineteenth century, parts of the production process have been mechanized in three phases (see Table 3.1).

Primary mechanization was aimed at speeding up and broadening the scale of goods and materials processing. This became possible due to the use of steampower and later the electric motor.

Table 3.1   Phases of mechanization

	Primary mechanization	*Secondary mechanization*	*Tertiary mechanization*
1850	Beginning		
1875			
1900	Spreading across sectors and maturing technically.	Beginning	
1925		Substantial diffusion in some sectors, increasing technical maturity.	
1950	Continuing but increasingly likely to occur together with secondary or tertiary mechanization.	Being generalized across a wider variety of industries.	Beginning in some industries but flexibility limited.
1975			Flexibility slowly increasing.

Source: Coombs (1984), p. 156.

Secondary mechanization is closely linked to the introduction of the conveyor belt, which was accompanied by Tayloristic control over manufacturing and transport. The employment of cybernetic modelling in the organization of production made a further mechanization of transport possible.

The development in computer technology has enabled further progress in primary and secondary mechanization. It has additionally provided a technological solution for the problem of control (*tertiary mechanization*). This has meant a new phase in the organization of production: from *manufacture* before the industrial revolution, via *machino-facture* in large parts of the twentieth century, to what might be called *systemo-facture*. The level of mechanization in the machino-facture era has not changed dramatically. In the new era of automation this alters rapidly.

A number of questions with regard to the changing structure of multinational corporations in the new era of systemo-facture can be posed:

—Does a process of centralization or decentralization of activities take place? Which activities become more centralized; which are decentralized?
—Do European multinationals make an effort to tap the creativity of 'intrapreneurs' on a large scale?
—How do companies make sure that efforts to automate activities of different divisions remain compatible with each other?

—Which research activities are carried out inside the firm, and which by outside contractors?

—What are the consequences of new inventory systems for the relations with subcontractors? There will be more subcontracting, but will there be more subcontractors? Will a more hierarchical (Japanese-like) structure of subcontractors come into being, or will the structure remain more similar to the less hierarchical American one?

As in Chapter Two, we first have a closer look at the European automotive industry. It will show how the introduction of new technologies affects the relationship with subcontractors. A crucial variable in this context is the percentage of value added as a share of total sales which is relatively high in the United States and rather low in Japan, with European companies somewhere in between.

3.1. IMPACT OF NEW TECHNOLOGIES ON THE STRUCTURE OF CAR MANUFACTURERS

The automobile firms were among the first to develop robots and other automation equipment primarily for their own use. But, until the beginning of the 1980s, rationalization in the automotive sector developed along the same incremental lines as in the preceding three decades. But then, a radically new production concept became feasible, based on technological innovations which were brought about in close cooperation with machine tool and electronic control machinery producers. In this concept fully automated painting and welding lines were employed. Final assembly is still mainly done manually, but third generation robots and vision systems will have a major impact on the level of automation of assembly activities in the future. Second generation industrial robots and (partly) automated inventory systems entered the car factories on a large scale in the early 1980s. Investment in up-to-date, i.e. highly automated, production equipment was linked to the introduction of new models and/or the installation of new production lines. Investments correspondingly increased enormously: the German car manufacturers, for instance, invested more than DM27 billion[1] in their production facilities in the 1980–3 period. International competition is not so much among new products—most of the cars increasingly look alike—but centres on

---

[1] See Jürgens, Dohse and Malsch (1986). Since the beginning of the 1980s the level of investments in tools and equipment has been at least tripled from DM1.9 billion in 1976 to DM7 billion in 1983.

production costs. Japanese manufacturers are said to produce small and medium sized cars at least $2000 cheaper than most of their US competitors.[2] From such a cheap production base it is easy to bear the extra cost of exports (transport costs, duties).[3] It is not surprising, therefore, that many producers have investigated the reasons behind this success. Large-scale automation, just-in-time or 'kanban' inventory systems, close links with subcontractors and quality circles are some of the central concepts in this discussion.

There are at least three features that give Japanese car producers a comparative advantage: the high diffusion of robots, the flexibility derived from the fact that Japanese car producers concentrate on assembly activities and receive most parts from subcontractors, and the higher degree of institutionalization of subcontractor relationships that helps Japanese companies to work with low inventories and to maintain high quality.

A first indication of the different nature of European (and American) producers compared to their Japanese competitors in the early 1980s was *the density of industrial robots per labourer*. In 1981 Toyota and Nissan, for instance, employed five to eight times as many robots as their US and European competitors.[4] Since then, however, most European firms have rapidly increased the number of robots installed, although they still seem to lag behind the Japanese.[5] FIAT advertises with the slogan 'Made by Robots', indicating that the firm thinks itself able to compete with the Japanese in price and quality head-on due to the high level of automation of production lines. As a result of fast automation the car industry in Europe and the United States became the *single leading user* of robots. In Table 3.2, data on the diffusion of robots within the plants of a number of major automobile multinationals are compiled.[6]

Only in Japan, the electronics sector has already bypassed the automotive sector in the use of industrial robots. With the further sophistication of assembly robots this is bound to take place in Europe as well.

[2] See, for instance, Tsurumi (1984).
[3] Even outright trade barriers have been insufficient to cope with the price difference. Most affected Western countries therefore have compelled the Japanese to 'voluntarily' restrain the number of cars exported to the EC and the USA.
[4] The robot density in 1981 for GM was: 0.0015; for Ford: 0.0012; for Volkswagen: 0.0018; for Renault: 0.0010; for Toyota: 0.0089; for Nissan even: 0.013. Malsch, Dohse and Jürgens (1985).
[5] PSA and Renault in 1984, for instance, had respectively 315 and 400 industrial robots installed in their French plants. This represents a robot density of 0.0016 and 0.0019 per labourer. (Direct correspondence with Association Française de Robotique Industrielle, 4 November 1985). Volkswagen in 1983 had more rapidly increased its robot density to about 0.044 per worker.
[6] In Chapter Five the long-term consequences of the concentration of robotics diffusion with the automotive multinationals is discussed.

Table 3.2 Diffusion of robots within plants of automobile multinationals, 1982/3

Company	Countries[a]						Totals[b]	
	USA	Japan	France	UK	FRG	Other	Installed	Predicted
GM	220				200	200[1]	2 400[2]	10 000[3]
FORD				120	116	120[1]	1 400[2]	6 000[4]
VW					521[5]			
DAIMLER					790			
NISSAN		125					1 100[2]	2 000[4]
BMW					200			
CHRYSLER						135[6]	240[7]	988[3]
AMC							58[7]	158[3]
MITSUBISHI		103						
TOYOTA		90					1 400[2]	
FIAT								
BRITISH LEYLAND								
VOLVO								
RENAULT			300					
PSA			301					
Share of car industry in national robot population		29%[8]	48%[9]	34%[8]	63%[8]	58% (Belgium)		

Sources:

ᵃ Malsch, Dohse and Jürgens (1984), Table 4, pp. 12–13.

ᵇ Predicted installations for 1988/90. See AMES/AFRI (1984), Annex 111, p. 180.

[1]Belgium; [2]1983; [3]1988; [4]1990; [5]diffusion in several sites; [6]Canada; [7]predictions; [8]1980; [9]estimate, 1983.

A second distinctive feature is the *value added* (as a percentage of total turnover). Firms which only assemble cars will have a lower value added than more vertically integrated firms. This is the case with Japanese producers. Toyota, for instance, which stands as a prototype of new production methods and of an extreme hierarchical type of subcontracting, is estimated to add less than 15 per cent to the final value of its end products,[7] whereas firms like DAIMLER–BENZ, VOLKSWAGEN or PEUGEOT in 1983 generated a value added of respectively 50, 41 and 39 per cent. Other European producers like RENAULT, FIAT and VOLVO are in a more or less intermediary position with respectively 31, 34 (in 1980) and 25 per cent value added. The Japanese firms thus are real assemblers. The European firms in general are still rather vertically integrated producers making most of the components themselves. But, the trend in European (and American) car manufacturing in the 1980s seems to become increasingly geared towards the Japanese example. A process of subcontracting parts of production takes place in order to concentrate on the design and assembly operations. This can be witnessed by a gradual shift in the relative weight of employment in car manufacturing firms in relation to employment in component producers or equipment suppliers like BOSCH–Europe or Lucas in the UK.[8]

Third, Japanese firms have a far more *institutionalized system of subcontracting*. Through financial participation in their most important subcontractors, they often have a direct influence on these firms which produce almost exclusively for them. The assemblers can order the needed parts on a day-to-day basis. A hierarchical system has evolved in Japan in which large subcontractors exist alongside and dependent on large assembly firms. The subcontractors themselves also have subcontractors. This system of subcontracting goes down to the level of outright home labour (see Figure 3.1, p. 93). Already in the 1960s, the final assemblers in Japan were demanding from their prime subcontractors a higher use of advanced (numerically controlled) machine tools,[9] which is one of the main reasons why numerically controlled machine tools and now robots are diffused much more widely in Japanese industry than in Europe or the USA.

Following the example of Japanese subcontracting, a number of Western automotive firms have started to institutionalize a similar relationship with their major suppliers in order to put out more of their operations. As in the Japanese case, they try to improve the production methods of their

---

[7] Other Japanese producers show comparable figures: Nissan (14.8 per cent), Toyo Kogyo (10.0 per cent), Daihatsu (6.5 per cent), Isuzu (11.3 per cent), Honda (9.9 per cent) in 1980. Chanaron and Banville (1985), p. 16.

[8] Chanaron and Banville (1985), pp. 16–19.

[9] See OECD (1984).

suppliers and at the same time to link their information system to their own system. A prime example is General Motors which in 1983 demanded its (electronics and machine tool) equipment suppliers to adopt its Manufacturing Automation Protocol (MAP).[10] Firms not willing or able to apply this protocol within a given period of time will be removed from the company's list of preferred suppliers.[11]

In an effort to institutionalize a more hierarchical subcontracting structure, French firms started a formalized group for the promotion of liaisons (GALIA, Groupement pour l'Amélioration des Liaisons dans l'Industrie Automobile) in 1984. The group aims to promote industrial standards in the French national context.[12] The group is selective in its choice of partners, which must have a viable technological and innovative capability. The effect could be the same as with the GM protocol: it conditions the subcontractors more to the needs of the main users; it will enforce a concentration in the number and size of major contractors (many small firms will not be able to cope with the high quality standards without the help of the larger user) and it will stimulate a quick diffusion of automation equipment, often in cooperation with the assembly firm. RENAULT, for instance, influenced one of its main subcontractors (Jaeger) to invest in a programmable flexible production line on the basis of the Japanese model and directly funded 15 per cent of the investment.[13]

In the 1970s FIAT adopted a pattern of putting out work from its Turin plant to local firms, artisans and outworkers. At the beginning of the 1980s FIAT declared that in addition to assembly work 'it will only produce the suspension system and technologically important parts of the car in-house'.[14] But at the same time as the volume of subcontracting increases, the number of subcontractors declines, because the assembling company can only assure the high quality of a limited number of suppliers, and a strategy of just-in-time deliveries spurs concentration among suppliers as well. One of the results is an enormous cut in the number of suppliers which declined by two thirds. The firm also actively

---

[10] MAP has been developed in cooperation with some of GM's major US suppliers like IBM, Digital Equipment, Allen Bradley, EDS (subsidiary of GM).

[11] GM tries to propagate the protocol to become a worldwide standard in CAD/CAM networks. Due to its sheer size and to its close cooperation with other companies in the MAP Users Association, the company will probably have considerable success in setting this *de facto* standard in factory automation.

[12] CPE (1985), pp. 30, 53. According to the same study a European initiative for organizing automotive firms and their subcontractors took shape in ODETTE, Organisation des Données Echangées par Télétransmission en Europe, in 1984. The producers of the four main countries and the organizations of equipment suppliers take an active part.

[13] CPE (1985), p. 57.

[14] Murray (1983), p. 92.

'encourages' the survivors 'to raise productivity and begin to sub-assemble parts in their own firms'.[15] Already 40 per cent of the Ritmo model is sub-assembled outside of FIAT's factories. The French and Italian firms can relatively easily adopt a more stringent subcontractor system because the share of components bought abroad is around or below 10 per cent. VOLKSWAGEN, which purchases around 20 per cent of its components abroad, or FORD–Germany (30 per cent) and SAAB (50 per cent)[16] will have more difficulties in organizing their subcontractors in a just-in-time mode.

In 1986 VOLVO announced a SEK12.5 billion investment programme out of which it finances a SEK7 million technical centre for subcontractors in the automotive sector.[17] AUSTIN ROVER in the 1982–5 period invested more than £100 million in computer-integrated manufacturing systems (closely cooperating with Honda in this respect). The result is an automated databank which directly controls all facets of the production process, not only from the first design to the assembly, but also the production of components by its subcontractors.[18]

In an international comparative study, under the coordination of the Massachusetts Institute of Technology (MIT), on *The Future of the Automobile* it is predicted that the reorganization of the car industry will 'lead to less formal vertical integration but closer operational coordination among members of the production chain, with *greater geographical concentration* of production at the point of final assembly'.[19]

In sum, the general characteristics of the industry[20] seem to point at a growing 'Japanization':

—*A smaller number of suppliers*, of which some will supply all firms, in order to realize maximum economies of scale and to adopt more stringent quality measures.[21] With the growing influence of electronics in production and products of the automotive suppliers, even traditionally strong subcontractors (e.g. BOSCH) can meet large newcomers on the market which try to diversify into the rather lucrative automotive components business from a strong production base in general electronics (for instance, OLIVETTI or SIEMENS). The decrease and concentration in the number of suppliers can pose considerable risks for the assembly firm. In 1984 the selective strike

---

[15] Murray (1983), p. 92.
[16] Figures taken from: P. Bianchi, *Components technology and the Auto Production Process in Europe*, Bologna: NOMISMA, 1983, p. 33; cited in CPE (1985), p. 55.
[17] *CPE Bulletin*, January 1986, p. 14.
[18] *Computable*, 7 March 1986.
[19] Altshuler *et al.* (1984), p. 189 (emphasis added).
[20] Combining also other tendencies as analysed in the MIT study.
[21] See also Office of Technology Assessment (1984a or b), p. 114.

strategy of German trade unions with strategic component suppliers like BOSCH affected large parts of the European automotive industry which consequently also had to shut down their factories (see Chapter Four). The car manufacturers in coalition with the central government in Germany are still in a legal struggle to decide who has to pay the wages of redundant workers in the large automotive firms.

—A *growing cooperation among final assemblers* in subsectors (for instance, engine production), in order to obtain efficient production units with enough scale economies.

—The *stimulation (or even enforcement) of automatization with primary suppliers* by the final assemblers. This increases the pace of technology transfer in process technology from the core films to their periphery.

3.2. INTERNAL REORGANIZATION

Many multinational firms outside the automative sector are simultaneously large users *and* producers of core technologies. For both a faster development and a better use of technologies, internal reorganization of the firm is necessary to take care of increasingly convergent technologies and to make extensive and rapid use of these same technologies in creating new efficient production structures. We will consider five different aspects of this reorganization: the consequences of converging technologies, the creation of independent business units, the centralization of automation, the restructuring of the R&D organization and the (de)centralization of decision-making structures.

3.2.1. The consequences of converging technologies

In coping with the growing convergence of technologies, especially in the area of microelectronics, firms have two basic options: either to *leave the organizational structure intact* and have the respective divisions all develop their own strategies of specialization or diversification into converging areas, or to *merge divisions* based on more and more identical technologies.

Some Japanese firms might provide an example for the *first* option. HITACHI, for instance, thoroughly restructured its operations and sees to it that factories producing mature products (with high market penetration rates) are looking for new activities, especially in microelectronics,

Table 3.3   Technological convergence, merging of divisions

ERICSSON	In 1983 creation of eight business areas in which, for instance, ERICSSON Information Systems combines both telecommunication and data processing and includes the acquired DATASAAB and FACIT companies.
STC	Takeover of ICL (computers); foundation of a separate 'networks' division in July 1985 (1000 employees) as part of the integration of both companies' activities. Supply of PABX, LANs and VANs mainly to large multinationals is foreseen.
PHILIPS	1985 merger of product divisions Telecommunication Systems and Data Systems. Public switching activities were located with the new joint venture AT&T–PHILIPS industries. In Germany the Dataprocessing and Cable divisions in 1981 merged into a new company called PHILIPS Kommunikations Industrie AG (PKI).
SIEMENS	1984 merger of Private Switching and Data-processing divisions into a single telecommunications group. End of 1983, creation of a Production, Automation and Automation Systems division with 10 000 employees of which 4000 in R&D, service, advisory and training, and with a total turnover in 1984/5 of DM2 billion.
THOMSON	1981/2 centralization of management of THOMSON CSF and Thomson–Brandt into one holding with central planning.
THORN–EMI	Reorganization of all its high technology subsidiaries into a single group (with six units ranging from defence systems, data processing, telecommunications to semiconductors) with a turnover of £700 million and a centralized management in 1986.

Sources: annual reports, *Financial Times*, *Business Week*, interviews.

and to a limited extent also in bioengineering.[22] But the divisional structure and the autonomy of the divisions in choosing their own technology path does not seem to be affected. For the firm's management this poses the problem of coordinating potentially overlapping activities.

European multinationals active in these areas have chosen the *second* option. Firms producing machine tools and data-processing equipment have often also opted for the merger variant. In merging divisions a more integrated strategy towards the factory and the office of the future can be adopted. Firms clearly hope to *minimalize interdivisional competition* and make *optimal use of technological synergism*. Table 3.3 inventorizes major examples in the 1980–5 period of large structural changes in a number of European multinationals.

The tendency of growing integration, however, is not without problems. Besides practical technological problems, it is clear that especially in the period *before* the organizational restructuring an intense competence

[22] Hitachi, *Annual Report 1984*, p. 6.

battle between divisions evolves, which can go on for a long time after the *de facto* restructuring has taken place.[23]

3.2.2. The creation of Independent Business Units (IBUs)

The creation of larger divisions, however, can lead to more bureau-cratization and delays in decision making, thus reducing the flexibility of a company to follow new developments. Some European companies have tried to solve the dilemma by stimulating 'intrapreneurs'. However, they do not go so far as IBM, for example, did by setting up real 'Independent Business Units' (IBUs).[24] European firms have probably anticipated some of the problems IBM has experienced such as the incompatibility of the personal computers with IBM's own mainframe computers as a consequence of a too rapid and independent development. This experience has prompted IBM to shift its attention back to its more centrally coordinated development laboratories and factories.[25] In the United States, high tech companies often give highly creative employees the staff and the means to follow up their ideas in an effort to keep talented people who otherwise might leave the company to set up their own enterprise. The chance that this would happen in Europe is much lower. Most European multinationals do not create IBUs, but have opted for a more controlled process of setting up 'task forces' instead, which cooperate for a specific period to develop and launch a new product.

Often, joint ventures with other companies have a similar function and can to some degree be compared with IBUs set up to keep intrapreneurs

[23] This happened with Ericsson, for example.

[24] IBM instituted more than ten IBUs in the 1980s. The Personal Computergroup in July 1980 was one of the first. Others are: production of hard disks for PC-AT, robots, software development (Milford, Conn., which is a very large unit), special computers and communication networks for universities, automized cash registers, diskette stations, computer systems for medical analyses (which died a silent death). David Sanger, in *Intermediair* (Amsterdam), 23 August 1985, pp. 1–9.

[25] Jeffery Hart notified us that there have been already major problems of incompatibility even across IBM mainframes when the IBM 360 was developed. IBM's solution to this has not been to insist on machine compatibility but rather to make their operating systems and higher level languages as similar as possible across machines, so that customers can easily transport their existing software from machine to machine. IBM has tried to create networking architectures by its SNA (Systems Network Architecture) efforts and more recently the introduction of a peer-to-peer token network system running under a software standard called LU 6.2. In its effort to become a major PC producer IBM used the better-designed microprocessor of Intel. A similar relationship developed with Microsoft Corporation, which is producing the operating system for the new PS/2 line of IBM PCs. IBM even bought a 20 per cent stake in Intel. IBM still uses Intel microprocessors in its personal computers, but it has used the time to develop also an own capability in this area. In 1987, IBM decided to sell most of its equity in Intel.

inside the firm. They certainly can lead to the same problems. The joint research centre (European Computer Industry Research Centre, ECRC) of ICL, BULL and SIEMENS in Munich is perhaps a good example of an IBU-like construction which combines the resources in basic research on machine-assisted decision-making processes and next generation computers.[26] A firm like ERICSSON, however, has explicitly chosen *not* to install Independent Business Units for fear of loosing the grip on its own activities and the compactness of its firm profile.[27] Although European firms experiment with IBUs or smaller task forces of intrapreneurs (see examples of SIEMENS and ICL in Table 3.3), these are mostly under the strict control of the central board. This is also the case with most joint ventures engaged in development and production of core technologies.

3.2.3. Centralization of automation activities

Advances in information technology provide companies with the opportunity to transfer part of their activity to independent suppliers while controlling their activity by integrating their planning into their own computerized information system. The next step is that this strategically important communication and computation system itself, the brain and nerve system of the corporation, can be transferred to a formally independent company. GENERAL MOTORS, for example, bought the company Electronic Data Systems (EDS) which will run the global communication network of GM and integrate CAD/CAM systems in one international database, build up MAP systems (Manufacturing Automation Protocol) to include data on the production capacity of different plants and machines in order to coordinate the use of automated equipment, and give shape to a worldwide management information system. Trade unionists in the German subsidiary Adam Opel AG claim that Opel's integration into this worldwide system will reduce the capacity of German workers considerably to influence management decisions.[28]

EDS has not only the task of running the international tele-communications and computation system, but also of rationalizing flows of information and production internationally by contributing to an optimal use of production capacity. In this way, decision making on crucial aspects is brought under a separate company (within the global

[26] Interview with ECRC managers, October 1985.
[27] Interview, Ericsson, November 1985.
[28] Interviews with Opel employees.

GM corporation) where it cannot be controlled any longer by workers in any of the GM subsidiaries.

Whereas in the case of GM, EDS and OPEL, the shift of activities from manufacturing subsidiaries to specialized service subsidiaries (EDS) takes place inside one international corporation, this is not necessarily the case with all firms. UNILEVER, for example, transferred its own internal computer and telecommunications system to EDS because the GM subsidiary has gathered considerable experience in this field and has developed software that can easily be applied to other companies as well. This would imply that very crucial activities which used to belong to the core facilities of a multinational corporation can be tranferred to outside companies, thereby increasing the decision-making power of management while at the same time reducing the possible influence of workers.[29]

Thus, although most European multinationals have been rather conservative in setting up independent units beyond their (divisional) firm structure, in at least one area they have not adopted this strategy at all. Many firms have created centralized automation departments, which some also brought outside the firm's official structure (either through joint ventures or as formally independent, wholly owned subsidiaries). Table 3.4 gives examples.

Relatively autonomous automation departments outside the firm structure have a number of consequences:

(a) They mean a centralization of influence over the automation policy of the entire firm. The division or the new firm is *directly* linked to the board of directors. Often the managing director of the automation department or firm has a seat on the central board. This indicates on the one hand that the power of the protagonists of automation has increased (they are no longer dispersed over several separate divisions, but can act as one block in the company with one clear interest). On the other hand, the control of central management over the overall automation process has increased too.
(b) The firm can *economize* on its own automation strategy. The selling of automation equipment and software to outside users makes it possible to earn back part of the investments the firm has to make.
(c) Taking the responsibility for automation away from the divisions deprives the employees of a direct information source on the company's automation strategy and thus *limits the power of labour unions* as mentioned in the example of EDS.
(d) Since the category of skilled and highly paid employees attracted to the automation departments in general is labour-union hostile, the *risk*

[29] Fitting (1985), p. 277.

Table 3.4   Creation of centralized automation units

SAINT-GOBAIN	Set-up of a new structure in 1982/3 of Competence Centres, the purpose of which is to develop robotics, Computer Assisted Design and Manufacturing (CAD/CAM), telecommunications, and automated office systems. In the field of CAD/CAM, all the group's technologies are involved. The Competence Centre for Robotics has grown rapidly. It now operates under its own corporate name of Sycpro and has opened its operations to companies outside the Group.
STET	In November 1984, a consortium was started: SEIAF of Electronica San Giorgio ELSAG, which supplies all kinds of control systems and coordinates the activities of the companies of the SELENIA–ELSAG Group in the field of factory automation (in joint cooperation with IBM–Italia), 'in order to promote the development of electronic and informatic systems in the field of factory automation' (i.e. FMS, CNC, CIM).
RENAULT	Merged its systems and automation divisions into one single legal entity, Renault Automation, integrating: SMC, Machine-outils Beauchamp (CAD), ACMA-Robotique, Evry Facility, SERI SOFERMO (CAD), SIRTES.
KRUPP	Founded KRUPP Datenverarbeitung GmbH as an independent company in January 1984. With a 40 per cent increase in sales to outside customers in 1984, almost half of total sales are to outsiders.
PHILIPS	TEO and CFT (former machine tool centre) departments used as automation units.
HOOGOVENS–DAF	Formation of Hoogovens Automation Systems (HAS) as an independent firm assisting in factory automation.
PSA	Set-up of PSA–AUTOMATIQUE INDUSTRIELLE as an external department.
FIAT	1981 creation of independent firm (COMAU) for production automation (1986 turnover of around $400 million).
GEC	In 1986 merging of Computer Aided Engineering and Automatic Test Equipment divisions which will closely cooperate with the Factory Automating Systems (FAST) division.

Sources: annual reports, interviews, etc.

of strikes in a vital area of the business strategy for the mother company *will decline* if it brings them together in one centre.

We find some of these factors also back in the reorganization of the R&D structure.

3.2.4. The restructuring of the research and development organization

Since new technologies move very fast and since it is difficult to assess beforehand which avenues of research lead to marketable products, many companies hesitate to engage their own 'in-house' research teams to explore radically new fields. Such a strategy would imply the risk that highly specialized staff attracted for the new tasks might eventually become obsolete if the initial research strategies prove to lead nowhere. In the 1980s large companies have been increasingly searching for possibilities to contract research with outside institutions — either specialized 'research boutiques', small venture capital firms, or with university institutes. Sometimes considerable overlap does exist between the different forms of contracting. Industry–university links seem to be very pronounced in the software area. This is not only due to the fact that this is a precarious field both scientifically and technologically, but also because there are 'at present no predominant industrial poles, so that research laboratories play a leading role here as they do not in micro-electronics nor computers'.[30] Since the integration of core technologies for a considerable part depends on the development of software it is not surprising that ICL, for instance, 'which is aiming for a more ambitious integration in the Factory-of-the-future than any other British company' has forged links with a number of universities and colleges since with regard to software 'there is more to be found outside industry than within it'.[31]

On the side of the participating universities this has raised a number of problems. First, 'faculty entrepreneurship' might lead to a change of priorities in the direction of easily commercialized research. Second, graduate students and postdoctoral fellows might be exploited by these faculty entrepreneurs for their own goals. On the other hand, however, universities are concerned about losing their senior scientists and hence have been reluctant to interfere seriously with consultative relationships or to prohibit faculty entrepreneurship. The advantages for the firms able to support universities can be considerable and there is abundant evidence that *all* major European multinationals of our sample are increasingly searching for links with research institutes.

If we consider the *effect of the new technologies on the organization* of research and development in the European multinationals one might be inclined to suggest that they could spur the decentralization of research activities over a number of internationally dispersed locations close to research centres of excellence. This has been done by some firms which

[30] *New Technology*, 4 February 1985.
[31] Op. Cit.

located research facilities in the United States (especially in biotechnology), although to a more limited extent than might be expected. A prominent reason for this might be the fact that domestic firms have fewer problems in cooperating with universities and research institutes funded by the national government than have foreign firms. In Japan this system is particularly closed. In the United States, growing concern with a possible flow of information to socialist countries also raised the barriers for Western European firms to use American university knowhow. European firms thus predominantly search for links with national (or, at best, European) research institutes, which limits the internationalization process of such vital company activity as R&D.

While more research is contracted out to specialized institutions, the remaining in-house research tends to become more centralized. This happens in spite of the fact that new telecommunications technologies (like teleconferencing) might also make more decentralization possible. There are a number of reasons why the dominant trend seems to be towards more centralization of R&D:

(a) The *increased interdisciplinary nature* of research due to the convergence of technologies and closer links with marketing (for design specifications) and procurement departments (for efficient use of subcontractors, see section 3.3) necessitates that different people from around the whole company work together. This will be the case especially in the core areas of a company's strategy.

(b) In forefront and interdisciplinary research *more expensive equipment* is used. This has to be carefully coordinated and efficiently used in central laboratories.

(c) A trend towards *common standards* (see Chapter Seven) and growing *integration of consumer markets* makes the centralization of specific research possible since no major national adaptation is needed.

(d) Since research becomes more critical for the survival of the firm in general, the *wish to control it becomes also more prominent*. The greater weight given to R&D has resulted in a higher ranking of the manager responsible for R&D in the central board of directors. In some firms (e.g. RHÔNE–POULENC) umbrella organizations are created to coordinate and control a coherent R&D strategy.

(e) Research becomes an activity where 'economies of scale' and 'critical mass' play an equally important role, as in production. Large teams of researchers are formed and in the area of software, where the distinction between research and production of the product is almost absent, 'software factories' must increase the *efficiency* of R&D 'production' by an almost Tayloristic division of labour.

In sum, a centralization of basic and other strategical important research seems to take place in big facilities (like the Bell labs of AT&T or the NatLab of PHILIPS) organizationally linked to the top of the company hierarchy, while applied research, development and adaptation to local circumstances can (and sometimes must) be carried out at a decentralized level.

Paul Wiltgen[32] comes to the conclusion that the degree of (de)central-ization of the R&D departments depends on the nature of their business and the degree of technological convergence in major areas of activity. In microelectronics, centralization and integration are more advanced than, for instance, in the chemical industry, partly due to the fact that in the microelectronics cluster product and process innovations are more closely related than in the chemical, food or steel industries.

3.2.5. Centralization of communication and decision-making structure

Firms increasingly make use of the advantages of telecommunications which make a decentralization of activities possible as well as a centralization of control. Essential for the relative autonomy of dispersed units is whether they can also communicate directly with each other, or only through the central computer/database. This is the difference between an 'all-channel' and a 'wheel' network of communication.[33]

Firms with more than one central office have used the new possibilities to link these offices to have them operate as one centralized unit by which they can save considerable amounts of money. ELF–AQUITAINE provides such an example. On the Paris–Pau (Pyrénées) trajectory, where its two major offices are located, it makes extensive use of the possibility of telephone meetings. Problems arise, however, when the central offices are in different countries as is the case with binational multinationals such as SHELL. The oil firm has not yet a very advanced internal communication network. From time to time difficulties arise in 'on-line' communication between UK and Dutch headquarters. Even between its UK and Dutch IBM computer systems compatibility problems can exist due to the US oriented standards of the equipment produced by IBM–UK and the European oriented standards of the products of IBM–Netherlands. Firms which try to link the internal communication system of major offices in a single country do not face these problems.

UNILEVER provides another example of a binational multinational having a very decentralized organization which uses telecommunications

---

[32] Wiltgen (1987).
[33] Williamson (1975), p. 46.

to integrate the company more. While the company in the past 'allowed considerable freedom for operating companies to use new information technology as they wished',[34] this will probably change as a result of the arrangement with EDS which has taken over the responsibility for the internal communication between UNILEVER affiliates.

We have not yet found examples of multinationals using telecommunications equipment to link their subsidiaries more with each other. Most firms rather seem to prefer to adopt a more or less centralistic structure in which a central database figures prominently.[35] Examples of European multinationals using a central database:

— ERICSSON has several IBM mainframe computers[36] in Stockholm which function as an on-line database for worldwide planning and information on research and production activities. Subsidiaries report all progress in research or insights in strategic planning to the central office.
— SIEMENS has centralized the management of the Megabit project with one member of the board of directors. He makes use of the knowledge of experts throughout the whole company through the use of a central computerized information system.
— BOSCH, through a data transmission network, has 'expanded the opportunities for increasing use of central computers by research and development departments from numerous locations'.[37]

Several general studies[38] only reach the cautious conclusion that the influence of the use of telecommunications on the decision-making structure is 'mixed'. We already noted, however, that if we consider the question of centralization versus decentralization in the context of overall business strategy, there is a tendency towards greater centralization of at least two vital functions of the new multinational corporation: the creation of more centralized R&D laboratories and the creation of independent (centralized) automation departments vital for the coordination of the restructuring process in multinational companies. Two other related aspects are: the increased efficiency in (central) financial management and the setting up of central databases. A decentralization of certain activities and a greater specialization among the operating units of the firms (for reasons

---

[34] Gillespie *et al.* (1984), p. 120.
[35] Some American companies form interesting examples of a centralized research structure. At the end of every shift in IBM's internationally dispersed research operations the day's output is transmitted back to the USA where it is collected; the sender units only receive 'need to know' information. Locksley (1985), p. 86. The design of new automobile models all around the world with Ford is centralized around a Cyber supercomputer posted in Dearborn (USA).
[36] Ericsson in March 1986 announced a cooperation agreement with Digital Equipment, which will result in the gradual substitution of IBM by Digital computers.
[37] Bosch, *Annual Report 1984*, p. 8.
[38] OECD (1983d); UNCTC (1984b).

of market presence, demands of minimal 'domestic content', spreading of risks, state procurement markets, use of local universities) thus has been accompanied by a *centralization of control* with most of the communication lines organized in a wheel-type network.

3.3. INSTITUTIONALIZATION OF A SUBCONTRACTOR STRUCTURE

The concentration on higher value added, investment and research intensive activities has increased the attention of most core firms on the organization of their subcontractor structure. At least two types of subcontractor structures can be discerned; see Figure 3.1.[39]

Most major multinationals seem to move towards the more hierarchical and formalized Japanese subcontractor structure.[40] They encounter considerable difficulties, however, for instance due to:

(a) *The dispersed nature of the producers' factories* PHILIPS, for instance, is clearly aware of the problem that in order to be able to adopt just-in-time inventory systems it has to centralize its production facilities. In several publications the firm has made clear that it is willing to follow the Japanese model in this sense.

(b) *National difficulties* Subcontracting on a just-in-time basis requires a very advanced system of high-quality deliveries. A very good example of the problems attached to this is provided by the experience of Japanese multinationals locating in the UK. Matsushita's television factory in Cardiff (Wales) had to cope with 30 per cent bad-quality deliveries which had to be returned. Compared with the return quote in Japan of below 1 per cent this is a devastating figure.[41] According to Ohmae, it will take the Japanese firms at least ten years to find ways to let their subcontractors produce at a lower deficiency level.

---

[39] More detailed (and less 'ideal type') treatment of the Japanese variant than that shown in Figure 3.1 can be found with Goto (1982) and McMillan (1984).

[40] According to a MTI study, the significance of subcontractors in all branches in Japan even increased. In the 1976–81 period the share of subcontractors in the group of small and medium sized firms shifted from 60.7 to 65.5 per cent. It concluded further that 'the number of sub-contractors employed in one enterprise in the past five years appears to be on the decline in most industries, indicating a clear shift toward the selection of a limited group of subcontractors. . . . As a result subcontractors with a good reputation in terms of their high level of technology and management skills receive the bulk of orders'. MITI (1984).

[41] The result in the UK was that the original domestic producers in the UK made bad products which prompted the customers not to buy television sets but to lease them. See Ohmae (1985), p. 201.

(a) The Japanese variant (ideal type)

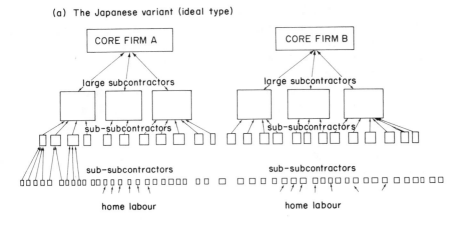

(b) The European and US variant

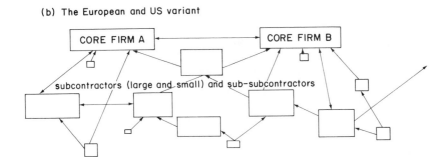

Figure 3.1 Two types of subcontractor structures

A more or less formalized strategy in helping subcontractors in order to have zero-deficiency input, enforce throughput time shortening of more than 50 per cent, adopt just-in-time inventories or one-day-trucking systems is provided by PHILIPS's *comakership project* (in the Netherlands paralleled also by large companies like Océ-v.d. Grinten, IBM, VOLVO-Car, Rank Xerox and DAF).[42] In 1984 PHILIPS started to approach 200 of its larger subcontractors in order to make pilot arrangements to analyse whether reliability of deliveries, product quality and shorter delivery times could be combined. The core company helps the comaker in the framework of a longer-term cooperation to adapt its production in an early phase more to the central firm's needs. PHILIPS considers the 'comakers' as an advanced

[42] Rank Xerox (Venray, Neth) adopted a 'Leadership Through Quality' programme aimed at higher quality and efficiency. Among other things, in 1984 it resulted in a reduction of the number of subcontractors from 4000 to 2000. A further reduction to not more than 500 firms is planned. *Financieel Dagblad* (Amsterdam), 22 May 1984; cited in Gaag, (1985), p. 91.

base for its own production.[43] PHILIPS has 22 000 subcontractors in the Netherlands alone. A sharp rationalization of the number and structure of sub-contractors will be one of the effects of the comakership project.[44] PHILIPS aims at only two key suppliers per category of supplies. The wider social impact is exemplified by one of PHILIPS's comakers which started to work for six days a week in order to be able to supply PHILIPS's needs.[45] The larger use of subcontractors and higher level of automation has lowered the value of unsold stocks and inventories from 28.9 per cent of total turnover in 1984 to 23.2 per cent in 1985.[46]

Comakership-like projects are adopted all around Europe by most large firms. They provide the large firms with needed flexibility, as in Japan. This flexibility, however, in fact is furnished by the small and medium sized subcontractors. A study by a Dutch bank (NMB) on the relation between large and small companies in industry comes to the conclusion that in the Netherlands and other European countries a hierarchical structure of large companies with (sub)subcontractors is evolving, representing many of the characteristics of the Japanese model.[47] This might also be an explanation for the fact that we have not found many large-scale examples of direct home-labour activities of the core multinationals in Europe.[48] Not the multinational companies themselves, but their sub-subcontractors (squeezed to supply at the lowest price possible), are inclined to make increasing use of home labour.[49] Interestingly enough, a comparable hierarchical structure was not found in the United States. With their economic structure the European countries thus occupy an intermediary position between the USA and Japan, but are becoming increasingly apt for 'Japanization'.

3.4. INCREASED CAPITAL INTENSITY AND DECLINING LABOUR CONTENT

A logical consequence of the strategies of most European multinationals towards more automation and more subcontracting would be that capital

---

[43] Gaag (1985), p. 75.

[44] An example: with the Philips–Drachten factory, 25 potential comakers were assigned in mid 1985. In the first year of the project the number of local subcontractors more than halved from 130 to 60. Gaag (1985), p. 74.

[45] Marcel Metze, Noord-Brabant, in *Intermediair*, 14 February 1986.

[46] Measured as percentage of total turnover. *Volkskrant*, 27 February 1986.

[47] Nederlandse Middenstandsbank NV (1985).

[48] One of the rare studies which mentions some examples of large firms using direct home labour is Steinbrinck (1985). According to the study, especially large US firms in the United Kingdom (IBM, Hewlett–Packard, Rank Xerox, Digital) make use of home labourers for low-skilled jobs.

[49] Läpple *et al.* (1986).

Table 3.5   Declining labour content in total output (millions of national currency)

	1975			1984		
	Expenditure on Personnel	Sales	%	Expenditure on Personnel	Sales	%
HOECHST	5 651	20 776	27.2	11 095	41 547	26.8
THOMSON–BRANDT	460	2 785	16.5	830.9	6 107	13.6[1]
RHÔNE–POULENC	6 240	17 875	34.9	13 148	51 207	25.7
SAINT-GOBAIN	7 121	21 164	33.6	18 100	57 894	31.3
AEG	5 604	14 097	39.7[2]	3 792	11 015	34.4
AKZO	3 109	9 717	32.0	4 292	16 520	26.0
BASF	2 003	18 801	11.1	6 190	40 400	15.3
BMW	1 439	5 959	24.1[3]	2 793	12 932	21.6
ICL	189	419	45.1[4]	452	1 124	40.2
MANNESMANN	3 345	13 094	25.5	5 715.	15 766	36.2
DSM			13.5			8.2
HOFFMANN–LA ROCHE	1 544	4 755	32.4	2 844	8 267	34.4
BP	95.4	7 781	1.2[6]	2 665	37 933	7.0
CIBA–GEIGY	2 997	9 035	33.1	4 893	17 474	28.0
BOSCH GROUP	2 772	7 281	38.1	6 563	18 373	35.7
KRUPP GROUP	3 610	12 746	28.3[7]	3 666	18 474	19.8[8]
GIST–BROCADES	226	863	26.2	444	1 747	25.4
SIEMENS	9 843	20 676	47.6	18 728	45 891	40.8
NIXDORF	267	616	41.7	1 227	3 273	37.5
RENAULT	5 581	18 264	30.6	15 633	72 105	21.7
ICI	1 227	8 256	14.9	1 743[5]	9 909	17.6
UNILEVER	6 699	36 705	18.2	11 701	66 791	17.5
OLIVETTI	225	856	26.3	539	2 552	21.1
DEGUSSA	538	3 127	17.2	1 082	11 122	9.7
SHELL	1 311	12 705	10.3	3 626	63 542	5.7
DAIMLER–BENZ	4 632	16 424	28.2	11 598	43 505	26.6
FIAT	1 016	3 835	26.5	4 240˙	20 000	21.2
VOLVO	3 613	13 692	26.4	7 432	87 052	8.5
VOLKSWAGEN	5 550	18 857	29.4	13 227	45 671	29.0
ERICSSON GROUP	584	1 673	34.9	7 856	29 378	26.7
ELF-AQUITAINE	969	34 500	2.8	7 213	134 000	5.3
ASEA	590	1 787	33.0	6 109	36 100	16.9
BRITISH TELECOM	2 394	5 708	41.9	2 715	6 876	39.5
MATRA	306	1 064	28.7	1 442	5 756	25.0
PSA	3 717	14 654	25.4	n.a.	n.a.	
THORN–EMI			28.6	841	3 204	26.2
BAYER	2 394	7 955	30.1	4 006	16 215	24.7
PHILIPS	11 212	27 115	41.3	n.a.	n.a.	

[1]1980. [2]1977. [3]1978: 'Personalaufwand in % der Gesamtleistung': 1975: 27.9%; 1983: 21.4%; [4]1983. [5]In the UK. [6]After 24.6% in 1983; [7]Krupp domestic in 1975: 27.6%; 1979 (group consolidated then). [8]Biased by oil business, correction would give 7383/59 315 = 12.4%.

intensity of their business increases and that labour content decreases considerably. Since large-scale automation has evolved earlier in the process industry, most of the multinationals in this field already have a very low share of wage costs as a percentage of total costs. Measured as part of total sales SHELL, ICL, BASF, DEGUSSA, ELF-AQUITAINE and BP pay below 15 per cent of their income on wages. Japanese firms like Nissan and Toyota in discrete (automotive) production already show much lower rates (7 and 8 per cent respectively) in their total expenditure on wages.[50]

From Table 3.5 it becomes clear that almost all European producers engaged in discrete production considerably lowered expenditures on wages as a percentage of sales in the 1975–84 period. They do not yet reach Japanese levels of capital intensity, but the trend is clear. A contrasting development, however, takes place in the process industry which tends to become a little less capital intensive: BASF, BP, ICI and ELF showed some increase in the share of wages. This can be an indication of their concentration on higher value added, higher R&D intensity and fewer bulk products, which in the short run can raise the total wage sum and lowers the income derived from bulk products' sales.

This chapter and the previous chapter have shown that European multinationals tend to become assembly firms more than integrated manufacturers. Large firms increasingly look like 'systems houses' which establish market leads in certain products and services, but to a considerable extent rely on outside suppliers of components. We would like to call this the 'Japanization' of multinational business.

---

[50] See Chanaron and Banville (1985).

Chapter Four

Social Impact of New Technologies

If large companies become research, design, assembling, trade and service organizations, what will happen to employment? If these companies spend less on wages and more on external supplies, does this imply an overall reduction of employment? Or does this only mean that employment shifts to the subcontractors, where more of the production takes place?

The fact that the share of wages in total sales has declined considerably in recent years (see Table 3.5) is the outcome of two different but overlapping trends. On the one hand, the introduction of new technologies will in most cases reduce the amount of living labour for those activities that are increasingly automated. This is the main driving force behind the introduction of these technologies. Its impact on overall employment, however, may be overstated if we do not take the new division of labour between central firms and their subcontractors into account. If the most labour intensive activities have simply been taken over by smaller companies, the effect is less dramatic than it otherwise would seem. But these activities are hardly ever transferred from large to small companies without being altered at the same time. The new division of labour between a large company and its supplier implies in many cases that the large company helps (sometimes even forces) the subcontractor to modernize, rationalize and partly automate its production as well—with the effect that although subcontracting increases in volume, the amount of labour necessary to produce the supplies needed may nevertheless decline. An overall increase in employment can then only be expected from generally higher economic growth, stimulated by the ongoing restructuring process.

The debate on the social impact of new technologies has largely centred around the question whether the application of new technologies will finally create or destroy employment. The different arguments are discussed in this chapter, which will come to a rather pessimistic conclusion (section 4.1). Directly related to the question of the amount of employment are the questions what the *qualification* of the remaining jobs is likely to be (section 4.2), and what the regional distribution of economic activities will be (section 4.3). Finally, we shall tackle the question how workers' influence on company decision making is affected by the application of new technologies in the production process and the resulting organizational changes described in Chapter Three (section 4.4).

4.1. THE IMPACT ON EMPLOYMENT

The discussion on the employment impact of new technologies remains highly speculative. Many predictions made during the 1970s concerning the period up to 1985 have obviously been blatantly wrong, some underestimating, others overestimating the degree of unemployment actually reached in Western European countries. But even seen in retrospect, it is very difficult to see *why* these studies have been wrong and how much of the actual unemployment has to be attributed to technological change or is 'simply' due to the economic recession — and how these two eventually are interrelated. In order to get a clearer insight, the German Ministry of Research and Technology (BMFT) in 1985 ordered a kind of 'meta-study' to find out why different analyses of the effect of new technologies on employment come to opposite conclusions.

4.1.1. The 'Meta-Studie' of the German Ministry of Research and Technology

The authors of this study[1] have compared studies predicting a high loss of employment with those that arrive at more optimistic conclusions. Negative consequences were expected in studies which concentrate on the *direct* effects of introducing new technologies. These studies usually analyse the *short-term* effects, treat the *micro-level* (company or department) and focus above all on *large* companies and on the automation of *standardized* production processes. Positive effects on employment are predicted by studies which include also *indirect* and long-term effects. These studies focus more on the *macro-level* (national economy). Besides large companies, *smaller* enterprises and automation of *non-standardized* processes are examined. The optimistic view that new technologies eventually will create rather than destroy employment is based upon this kind of study. A further distinction can be made between the *product* and *process* dimension of new technologies. Emphasis on product innovations, which can create new employment, leads to an optimistic view. A concentration on process innovations which accelerate the rationalization process implies a more pessimistic view on new technologies.

[1] Friedrich and Rönning (1985).

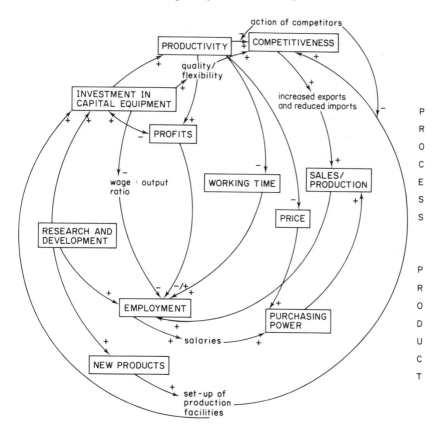

Figure 4.1   Product, process innovation and employment: contradictory effects

Figure 4.1[2] tries to summarize main elements of major assessments. One line of reasoning is that investment in capital equipment leads to productivity gains (including also higher quality and levels of flexibility), which can raise competitiveness. Assuming no similar action of the main competitors, this makes more export and/or a recapturing of parts of the home market possible. Total sales rise, with positive effects on employment. Employment increases further as a result of the multiplier effect created by the demand stemming from additional wages payments.

Investment in capital equipment, however, also lowers the wage: output ratio and has a squeezing effect on profits which influence employment negatively.

[2] For comparable models see Boyer and Petit (1980); Jeandon and Zarader (1983), p. 189.

On the product side, new products can create new markets and thus enhance the competitive position with a positive impact on employment. In principle *public policies* would have to be added to the model since they affect *all* sequences of the model through subsidies, procurement, labour regulations, protectionistic policies and the like.

In this section some of the above contradictions will be elaborated. Five apparent 'antitheses' will be dealt with: direct versus indirect impact (4.1.2), process versus product innovation (4.1.3), micro versus macro impact (4.1.4), large versus small companies (4.1.5) and short-run versus long-run impact on employment (4.1.6). This analysis leads to the conclusion that the pessimistic view in the short run (i.e. until about 1990) has to be enunciated due to the fact that the diffusion of new technologies in the economy is a slower process than often predicted. The optimistic view, however, is *unrealistic* even if indirect and long-term effects, product innovation, macroeconomic coherence and jobs created by small firms are taken into account. The employment problem to which European multinationals contribute will increase instead of decrease.

4.1.2. Direct versus indirect effects

New production processes do not only destroy jobs, but new technologies also create new employment, for instance jobs needed for the design, production, installation and use of computers, robots and numerically controlled machines. It is often said that employment patterns are only shifting: simple and easy-to-automate jobs will be replaced by more responsible jobs (installing and operating the new machines). But can this new employment compensate for job losses? The situation in the USA has been taken as an example for European policy makers. *Employment in the USA did not decline* in the 1980s. Some two million jobs were lost but, at the same time, 4 million new jobs created. Which kind of new positions were created? Did new technologies indeed create as many jobs as were destroyed in other fields?

The US Office of Labour Statistics issued a report based on the experience of this period, in which the expected growth of labour between 1982 and 1995 in different occupations is extrapolated.[3] Technology-related professions do indeed grow fastest. The number of computer specialists will almost double until 1995. Demand for systems analysts, programmers and computer operators will increase strongly as well. These jobs, however, represent only a small percentage of the US labour force:

[3] OECD Labour Force Statistics.

the fast growth therefore leads only to a comparatively small growth in the number of jobs. Other sectors are responsible for the highest absolute increase in level of employment. On top are porters and security guards. Between now and 1995 more doorkeepers, night-watchmen and porters will become employed than computer technicians, systems analysts and program engineers together. New cashiers as well as new secretaries will outnumber the new computer personnel needed. Low-skill professions do not grow as fast as the new technology-related jobs, but they create more employment in absolute terms because they represent a much bigger share of the total labour force.

A closer look at developments in the United States shows that the mass of new jobs that are created there are in the field of simple services. Given the different situation in Europe (still fewer security problems in the cities, less need for fast-food restaurants), this example cannot easily be followed. Only relatively few new jobs have been created in the manufacturing, installation and maintenance of the new means of production and in the new services that have become possible as the result of new technology. The often heard reference to recent developments in the United States cannot suffice as an answer to the question how employment will develop in Europe. The precarious balance between jobs lost through process innovation and jobs gained through product innovation has to be analysed in more detail.

4.1.3. The contradictory impact of product and process innovation

It is hardly disputed that the new production processes based on applications of microelectronics are labour saving. Their labour-saving impact will be illustrated below. What is highly disputed, however, is to what extent this reduction in employment is compensated for by increasing employment in the industries that manufacture and service the new devices. For this reason, we shall have a closer look at the employment created in Europe in those industries in which the new technologies are developed and manufactured. Then, we shall turn to the question whether the production of hitherto unknown products and services, made possible by the technological development, is able to restore the balance. Finally, we shall consider the macroeconomic argument that increased profits based on cost reduction and additional income for the producers of new equipment, new services and new consumer products, would lead to more overall demand and consequently higher growth. The additional growth, optimists believe, would create enough jobs to make the overall impact of new technologies on employment positive.

The impact of process innovation

Most studies dealing with the Factory-of-the-future, the Office-of-the-future or the Farm-of-the-future come to the conclusion that the new technologies have a significant direct labour-reducing effect. This is due to the substitution of numerous activities on the factory floor by robots and other automates, the replacement of much secretarial work by data-processing equipment, and the replacement of farm workers through an increasing use of automated machinery and new agrochemicals. 'Islands of automation' in the different fields of economic activity (like design, production, administration) can increasingly be linked to each other with the effect that a large number of activities that consist of transferring information from one sector to another, will become redundant. This involves activities such as passing design data on to the shop floor, adjusting machines to new specifications, counting inventories and filling in demands for the replenishment or the translation of sales figures into production planning.[4] The increasing use of CAD technology provides the database that facilitates full automation in all different sectors. The application of CAD/CAM helps to suppress the enormous research budgets, which will result in a declining number of researchers.[5] A decline in the *number of components* also decreases the amount of assembly work. This is exemplified by a drastic cut in the number of components in television sets,[6] or telex machines operating on the basis of one microprocessor replacing hundreds of other components.[7]

The higher precision and welding quality of robot-aided production makes it possible to reduce the number of welding spots in any metal-working assembly operation. New types of cars or other products which have to be welded, painted or assembled by robots nowadays are *designed beforehand* to be produced by automates with higher constant quality, thus having a cumulative effect on job displacement in all kinds of manufacturing industries. That robots replace labour is undisputed, it is only a matter of discussion to what extent.[8] The diffusion of word processors and other terminals in the office will in the medium term have even more impact on office jobs than robots on factory

---

[4] See OTA (1984b), p. 128.

[5] Siggraph exhibition in San Francisco, as reported in *Computable,* 13 September 1985.

[6] See example in Chapter One.

[7] See Evans (1984).

[8] Different studies come to assessments ranging from two to nine workers being replaced by one robot. An internal study of Philips TEO (Technical Efficiency and Organisation) department assesses a labour-saving effect of 20 per cent in industry due to the use of robots, whereas the OECD only reckons on a 1 per cent decrease of employment in manufacturing due to the use of robots.

work.[9] The increased use of automation in services like banking and insurance is also undisputed. In sum, a process of *industrialization of the service sector* and *informatization of industry* is taking place, which will result in a net loss of employment in all those industries *using* new technologies in their production. Highly integrated systems, however, will be employed only in the 1990s. The US Office of Technology Assessment expects for the US context that these systems will probably remain concentrated in the machinery and transportation equipment industries.[10] The real full-scale (negative) impact of process innovations on employment will therefore only be felt over a decade and depends very much on the speed of progress in areas like sensor technology, and the pioneering role of large enterprises (see later in this chapter).

Job creation through product innovation?

While it is rather obvious that process innovation results in a direct loss of employment, it can be asked whether this loss is not compensated for by product innovation. The new machines that are employed in process innovation are new products themselves that have to be designed and produced. This might create additional jobs. To get a better idea of the employment-creating effect of the production of new equipment, we shall have a short look at robotics, telecommunications equipment and production, before going into the employment impact of product innovation in general.

In *robotics*, the two European countries with a considerable domestic industry, besides Sweden, are Germany and France. In *Germany* the value of robot production grew from DM125 million in 1980 to DM410 million in 1984, whereas employment in this industry grew from 12 000 to 18 000 in the same period.[11] A threefold growth of production, thus, was accompanied by an employment growth of 50 per cent. Productivity in the industrial robot industry firms has doubled. In *France* employment in the robotics sector grew from 3900 in 1980 to 5100 in 1983,[12] i.e. by 30 per cent. There are no exact figures on sales volume or value. The number of production workers in the French robotics industry, however, has relatively declined in the 1980-3 period in favour of employment in research and marketing. This is also an indication of a considerable and

---

[9] Commission of the European Communities, DG-V, October 1984, p. 38.
[10] OTA (1984b), p. 135.
[11] Bundesministerium für Wirtschaft (1985), pp. 3–4.
[12] Chenard and Pino (1984), pp. 79–81.

quick growth in productivity. The robotics industry is obviously one of the first employers of its own devices — not just for the rationalization of production, but also to increase its knowledge of application problems and to install demonstration plants for marketing purposes. Employment creation as a result of increases in robot production therefore remains rather limited.

An example of successful product innovation is the ASEA Group of Sweden with regard to robotics. In 1972 a handful of people built a prototype robot. In 1984 the production of robots provided 1300 people with a job.[13] The link with process innovation becomes clear if we relate the expansion of the industrial equipment group of which the robotics section is part with overall employment developments in the other divisions of ASEA. First, the productivity level of this division grew rapidly with robots producing other robots (from 1981 to 1983 invoiced sales grew by 70.6 per cent, whereas employment grew by 24.5 per cent).[14] In all divisions, ASEA has fiercely restructured its operations in the 1980s and decreased employment, making use of the product and process technology developed in its own industrial equipment group. The effect on overall employment has therefore almost been neutralized. In the 1980-4 period, as a result, employment with ASEA has very slowly increased, with rapidly growing investment in capital equipment, lower wage:output ratios, and a tripling of sales. If we thus use these figures in the model presented in Figure 4.1, it becomes clear that despite the very positive influence in the case of ASEA of product innovation, level of sales, competitive position, R&D expenditures (growing to 8 per cent of turnover) on employment, this is *almost completely offset by the negative influences on employment* especially due to investment in process innovations. In firms with lower product innovation potential and a weaker competitive position (ASEA seems a positive exception in the European context) the overall effect certainly tends to be negative.

In the *telecommunications equipment* industry even the statistics for the USA show a decline in employment after 1979. Before that time, employment was growing.[15] A parallel pattern can be observed for the major European telecommunications equipment producers where a small average increase in employment in the second half of the 1970s was completely undone by a quick reduction in the number of employees in the 1980s. Only ERICSSON, which is 77 per cent dependent on telecommunications activities, in this respect shows an anticyclical pattern, but it started to shed labour on a large scale in 1986.

[13] Speech of Percy Barnevik, IRM/FT conference, Munich, April 1985.
[14] ASEA, Annual Report 1983.
[15] US Industrial Outlook, 1983 Survey of Current Business.

Table 4.1   *Increase/decrease in employment due to introduction of microelectronics*

	Products	Processes
Germany	− 14 000	− 27 000
France	+ 3 000	− 14 000
United Kingdom	− 6 000	− 28 000

The weak position of European multinationals in *chip* production has resulted in only a very small number of jobs in this industry relative to the USA and Japan (even within the large companies). The same is true for the European *computer* industry. Evans notes that between 1974 and 1982 employment in the EC in data-processing machinery and office equipment fell by 28 000 (i.e. 10.6 per cent). In the same period employment in electrical engineering in the EC fell by 338 000 (i.e. 14.2 per cent).

In general, every new product that is successfully introduced will of course create some employment in producing it. Many products now produced by European multinationals did not exist ten or twenty years ago. Many of these products, however, directly *substitute* older products which reached maturity. The production of large radios was partly substituted by small portable versions and later by hifi systems and the walkman. New generations of memory chips directly replace previous, less integrated versions. The same production personnel, therefore, can make new products, without any impact on the level of employment.

The impact may even be negative, since the new products often contain electronic components that can substitute hundreds of moving parts and can easily (often automatically) be produced and assembled (see 1.3.1).

A comparative study on the employment effects of the introduction of microelectronics in three European countries (France, Germany and the UK) in 1982/3 confirms this impression (Table 4.1).[16] Only the French industry in this sample registered a positive impact on employment due to the introduction of microelectronics in its products, probably because the French industry at the same time also had the lowest overall number of products integrating microelectronics. In general, the impact on employment of microelectronics in European industry, whether looking at the product or process side, is negative.

The hope that new products will create employment, therefore, is very much based on hitherto unknown products that will answer new needs and thus not substitute existing production. The argument is a very speculative one and cannot be proved or discarded in an empirical way.

---

[16] VDI-Technologiezentrum Informationstechnik (1985), p. 21.

Many of the differences in expectations with regard to future employment have their roots in whether different authors share this hope or not.

Product and process innovation become increasingly interlinked. Product design changes in order to facilitate automation of production, and new production processes make new products possible. It is the interaction between product and process technology that leads to negative expectations with regard to future employment, even if many totally new products will come up.

4.1.4. Microeconomic versus macroeconomic perspective

The optimism of those studies that predict little labour displacement is mainly due to their macroeconomic level of analysis. The development of new products and the cheaper production of existing products, the argument goes, will result in new demand which in turn will lead to economic growth and new employment.

A limited model, based on specific assumptions with regard to demand elasticity, easily leads to such a result. If we, however, include the problem of *demand financing* (see Figure 4.1) another picture emerges. People who are unemployed, or feel threatened by unemployment, will reduce their purchases. This influences the general level of demand. In most OECD countries the link between productivity growth and the growth of purchasing power, one of the basic prerequisites for the Western growth model, has loosened: productivity now increases faster than purchasing power. Economies facing this disequilibrium are increasingly inclined to export more in order to make use of the purchasing power abroad.

European countries are already far more intertwined with the world market than the USA and Japan. Lower prices can increase an industry's competitiveness and as a result also exports. But measures to increase productivity in other countries continuously threaten any industry's market share. All countries collectively cannot improve their trade balance at the same time. Parallel improvement of productivity in many countries only increases the gap between productivity growth and purchasing power in the Western world taken together.

It therefore becomes increasingly questionable whether technological development can provoke a lasting export boom in a world economy which is worried by *growing protectionism* of the industrial countries and by huge *debt problems* of Third World countries. Everywhere the same arguments are used. All countries adopt export oriented strategies. However, if the domestic purchasing power in these countries does not grow, it becomes harder to increase mutual exports. A macroeconomic analysis, which includes also the demand side, the financial dimension and a global perspective, hardly can result in a more optimistic result than an analysis at the national (macroeconomic) level.

4.1.5. Distinction between large and small firms

Analyses predicting high labour displacement are said to be based on the experience in large companies exclusively. In general, large firms will introduce automation more quickly in all parts of their operations. The Office of Technology Assessment notes:

> Clerical employment is most likely to change, at least in the near term, among larger firms because they are quicker to adopt computerized inventory and planning systems, and because they have greater information flow needs and problems. Larger firms are also more likely to adopt sophisticated automated materials handling systems . . . which are most economic in larger installations. Finally, reductions in company work forces as well as automated recordkeeping may affect personnel and payroll clerk employment.[17]

Labour-displacing effects of robots are much greater when applied in integrated production chains than as 'stand-alone' items in small-scale operations.[18]

In Europe, the multinationals as a group still increased employment in the second half of the 1970s, while the average employment in European manufacturing declined by 1 per cent.[19] But during the next five-year period (1979–84) multinationals cut employment much more than the average European decline in employment of 2.4 per cent.[20] This does not seem to be a cyclical phenomenon only. In the future, the direct contribution of multinationals to employment will probably remain limited, because of the quick pace of process innovation, the trend towards a concentration on research on the one hand and assembly operations on the other, and increasing subcontracting and flexibilization of the workforce (see Chapter Three). Any study focusing exclusively

---

[17] OTA (1984b), p. 142.
[18] Industriegewerkschaft Metall (1983), p. 35.
[19] Mid year estimate; OECD Labour Force Statistics.
[20] An exception to this general pattern in the 1974–9 period was formed by all major Dutch multinationals and two out of three Swedish multinationals. These firms had already reduced their employment much earlier, while increasing labour productivity. This may be indicative of having to operate in a small country, which forces them to automate more aggressively in order to remain competitive in world markets.

on employment in the large companies, therefore, would indeed overestimate the negative employment impact of the application of new technologies.

The optimism based on the expected employment creation by small and medium sized firms, however, should not be exaggerated either. Increased subcontracting will not necessarily lead to more employment with the subcontractors, for at least two reasons.

First, diffusion of new technologies among small and medium sized enterprises often is stimulated by the larger companies. IBM, for instance, lends capital equipment to its small suppliers not able to invest in this equipment in order to help them to produce the required quality products for IBM. In Chapter Three other examples were given. These suppliers will have a firm incentive to automate quickly in conjunction with their large procurer. This will diffuse automation technologies more quickly. This process has been especially pronounced in the automotive industry.[21]

Second, there is the tendency of the larger firms to use fewer suppliers (with higher quality standards). A concentration process in suppliers will, of course, have a considerable impact on their employment potential. The software business, for instance, which at the moment consists of thousands of small and medium sized companies, is expected to concentrate very rapidly and to become more automated, thus reducing its impact on employment creation.

4.1.6. Short versus long term

Will an assessment of the long-term impact of new technologies on employment come to more positive findings than an assessment of short-term impacts? This is the assumption spelled out in those studies that expect a positive contribution to employment from the economic growth fuelled by the higher incomes generated by increases in productivity. In our view, however, the short-term effects, on the contrary, will be less negative than the impact in the long run. As indicated earlier, the short-term effects of high technologies have been less pronounced than often feared. This is easily explained: when process innovation is introduced, initially a mixture of old and new machines is used. It takes some time to retrain the personnel and to programme and install the new machines. Company logistics have to be adapted. In such *transition periods* new technologies even cause

---

[21] See OTA (1984b), p. 114; a study of the OECD (1984) comes to the conclusion that early stimulation of quality standards by the Japanese automobile manufacturers, enhanced the rapid diffusion of numerically controlled machine tools with small and medium sized subcontractors.

additional need for personnel. In the highly automated Renault factory at Douia, for instance, 241 extra jobs were necessary in 1983.[22]

But as soon as the automated factory has overcome its teething troubles, the personnel can be reduced (in the Renault example this decline started in 1984).

The same process can be observed in the German automotive industry in which a modest growth of employment in the first half of the 1980s is expected to be the prelude to high redundancies in the second half of the decade.[23]

The *moment* of dismissal of labour affected by automation is not necessarily fixed to the installation of new machines, but more to the development of company sales. As long as sales are satisfactory companies can retrain workers and re-employ them in other parts of the production process. Only when sales stagnate will workers be dismissed whose jobs had already been subject to technological rationalization in an earlier period.

Part of the employment decline that results from the application of new technologies thus only shows up at a later stage. Part of the employment initially created in this process may itself be lost again in the long run. Many of the 'promising' jobs created in the information technology industry are also subject to rationalization. These include occupations like card-punchers, programmers and robot operators, which are made redundant by new forms of data input, more user-friendly operating systems that enable laymen without any knowledge of programming languages to use a computer, and a higher integration of CAD/CAM systems.

The long-term prospects are even more bleak, if a kind of *quantum leap in automation* takes place, which many observers expect for the years to come. A number of reasons can be put forward which corroborate these expectations:

(a) the advance from 'intrasectoral' to 'intersectoral' automation,
(b) declining prices for automation equipment,
(c) the coming to an end of the 'learning period' with regard to the application of the new equipment,
(d) the backlog of investment because of the hitherto unfavourable business climate,
(e) changing power and attitudes of trade unions,
(f) more active government promotion of the diffusion of new production processes.

(a) *Integrated automation systems* Automation in the past has often been either 'island automation' with the introduction of 'stand-alone'

---

[22] See Santilli and du Tetre, in AMES/AFRI (1984), p. 226.
[23] See Jürgens, Dohse and Malsch (1986).

robots, word processors or CAD terminals, or else remained confined to one of the three main spheres of activity of any company: design and development, production, and administration. The new equipment increasingly allows for *intersectoral* automation that will link CAD equipment, computer-integrated manufacturing and administration, thus making many jobs redundant that consist mainly of passing on information from one of these spheres to another. This opens up tremendous potentials for further rationalization.

(b) *Declining prices*   Many managers have hesitated to introduce the latest process technology, because they expected a similar development of price performance ratios as with computers, wherever more powerful systems have become available at a fraction of the price of a few years ago. Some also wanted to wait until some technical bottlenecks (e.g. with vision systems and tactile sensors) would be solved, which by and large might soon be the case.

(c) *End of a 'learning period'*   Instead of embarking on large-scale investment in new automation processes, firms have invested in a few pieces of equipment only in order to gather experience first. Research in Sweden on the extent to which firms make a cost–benefit analysis before they invest in robots has shown that only half of the (ten) firms investigated had based their investment decision on such an analysis.[24] The other firms had seen it more as a strategic 'investment in learning' to acquire knowhow that they would need in any case.

VOLKSWAGEN, striving to experiment with the further automation of car production, seems to adopt a strategy of automating different parts of its automobile production in *different pilot sites*. In Wolfsburg the production of the front part of cars is automated, in Emden the production of the back part and in Ingolstadt (Audi factory) production and assembly of the car's roof. At all these sites, the consequences of automation will still remain limited. But the company in this way accumulates the knowhow necessary to automate *all* stages of the production process. The integrated application of this knowhow in future production lines will have a more far-reaching effect on employment and work content than the sum of its parts.[25]

For many companies, the 'learning period' has now come to an end. They have figured out which kind of equipment will fit their needs best. They have trained people to operate and service the new equipment and got their workforce used to working alongside robots. They seem ready to invest on a larger scale in automation equipment — and they are forced to do so because the learning period of their competitors has come to an end as well.

---

[24] Edquist and Jacobson (1984), p. 6.
[25] See Malsch, Dohse and Jürgens (1984), pp. 36–40.

(d) *Backlog of investment* In the early 1980s, technological development proceeded, but many companies were hesitant to invest in new equipment because of a stagnating demand for their products. With underutilized capacities, they did not see any reason to invest in new machinery at all. Accordingly, a reservoir of new technologies has come into existence which wait for application. With a more favourable business outlook in the mid 1980s, companies seem more inclined to make use of the new technologies than a few years ago.

(e) *Changed position of trade unions* Large-scale investment in automated systems is often only made if the equipment can be used around the clock, all year long. This is only possible if the workforce is prepared to accept multishift work. Trade unions have opposed multishift work in the past, but their strength has eroded in many countries, and their attitude has sometimes changed as well. In some cases, multishift work is accepted as part and parcel of a bargain that includes a general reduction of working time. Thus, the opposition of trade unions that stood in the way of large-scale investment in automation has considerably weakened in recent years (see also below section 4.4).

(f) *Government support of automation* Finally, government action contributes to the expected quantum leap. While early government support of automation concentrated on the *development* of automation equipment, recently European governments have also taken measures to promote the *diffusion* of automation (see Chapter Six). The German Ministry of Research and Technology (BMFT), for example, expects that the result of its CAD/CAM programme will be an increased penetration of CAD technology in the German machine tool industry from only 4 per cent of all companies in 1984 to 60 per cent at the end of the programme in 1988. If this development really takes place, it certainly will mean an automation jump in German industry which will be relevant not only for the design phase. Since the introduction of CAD provides the opportunity to create a database in the early phase of a new development, which can be used in production and administration, too, the diffusion of CAD technology can have a far-reaching impact on automation in other areas as well.

4.1.7. Conclusion

The above analysis has shown that there is not much ground for an optimistic outlook regarding the impact of new technologies, even if

—not only the direct, but also the indirect consequences are taken into account,

—the potential for process *and* product innovation is looked at,
—an analysis at the macroeconomic level is added to the one at the microeconomic level of individual firms,
—not only large firms but also smaller firms are taken into consideration,
—and short-run as well as long-run implications are taken into account.

The outlook will be even worse if a quantum leap in automation does materialize, for which many preconditions now are given.

4.2. MULTINATIONALS, NEW TECHNOLOGIES AND QUALIFICATION

Not only the number of jobs is affected by the introduction of new technologies, but also the qualification of the workforce. While the impact on employment probably is negative, the impact on qualification seems to be positive in the sense that the remaining tasks demand a higher average qualification of the workforce than before, and this is especially the case for the employees of the large companies. Ever *higher skill levels* are needed to keep pace in the technology race. This is not only due to the increasingly complex nature of the production process which requires multiskilled workers able to operate, inspect and maintain the equipment. Larger R&D efforts also make a better-trained workforce mandatory, as product development and actual production are brought closer together. Producers of high technology products also increasingly have to provide external training for their suppliers and customers in order to keep a competitive edge in the race for provision of more system oriented and customized goods. The European multinationals thus provide lower total employment but tend to need higher-than-average job skills. As a result, for example, the ratio wage earners : salaried employees with SIEMENS altered in the 1965–85 period from 2.0:1 to 0.9:1.[26] The composition of the workforce with CGE in France showed a comparable change with a sharp decline in the share of 'workers' (from 61 per cent in 1971 to 47.6 per cent in 1984), while the share of technicians and engineers and executives increased (from 19.1 to 32.0 per cent in the same period).[27]

A more automated electronic components production site with

---

[26] Siemens, *Annual Report, 1984*. This is, of course, also the result of differing labels given to jobs in the process of negotiations with unions.
[27] CGE, *Annual Report, 1984*, p. 38.

THOMSON showed an overall decrease in personnel of 40 per cent. Unskilled workers, however, were more affected: their employment decreased by more than 76 per cent.[28] In the automobile sector employment prospects for the second half of the decade show a comparable pattern: RENAULT, for instance, will employ more technicians, engineers and professional workers but will reduce the number of *ouvriers de fabrication* and middle management (*agents de maîtrise*).[29] The German automobile industry knows fewer layers of middle management and consequently only assembly operations, which is the largest employment category and which requires much (semi)skilled labour, will be susceptible to employment reduction.[30] In the electronic components sector, which in Europe is dominated by multinationals, the share of qualified workers in total employment declined from 80 per cent when electro-mechanical components used to be produced to 35 per cent in the production of large-scale integrated circuits. At the same time the share of engineers and technicians and of non-qualified workers increased sharply (from 5 to 30 per cent and 15 to 35 per cent respectively).[31]

The still increasing proportion of qualified personnel makes it necessary for firms to *invest more in training and retraining*. For large companies a combination of in-house and external training can make it possible to economize on high training expenditures, whereas smaller firms mostly cannot provide their own training.

PHILIPS intends to increase expenditures on training of personnel even to 10 per cent of the wage sum after 1990 (it is now 2–3 per cent).[32] SIEMENS puts 'special emphasis on . . . providing continuing education to equip employees for the more demanding jobs generated by the introduction of new technologies and processes'. The company spent DM255 million in 1984 for continued education.[33] CGE's budget for training programmes is approximately 2.5 per cent of total wages (FF315 million in 1984).[34] At BULL, expenditures on training have already reached 7 per cent of the total payroll.[35] These efforts are not restricted to the electronics industry. At BAYER, for instance, more than a quarter of all employees participate in yearly training.[36]

The competitive strength of companies seems to depend increasingly on their internal capacity to adjust to new skills needed given the fact that

---

[28] Coriat (1984), p. 111.
[29] Richter, cited in Coriat (1984), p. 112.
[30] See Jürgens, Dohse and Malsch (1986), p. 9.
[31] See ETUI (1982), p. 20.
[32] Interview, Philips.
[33] Siemens, *Annual Report, 1984*.
[34] CGE, *Annual Report, 1984*.
[35] Bull, *Annual Report, 1984*, p. 38.
[36] Reply to questionnaire.

the general educational system is much slower to adapt to changing needs. This is not necessarily limited to the higher echelons of the workforce. It has been said with regard to the French automobile sector, for instance, that the high number of migrant workers severely hampers a process of quick restructuring, because these workers are often illiterate and can therefore only be retrained with considerable difficulties.[37]

The quest for higher skills by the large firms not only affects internal training, but also the companies' relationship with outside institutions. They have become more aggressive in their acquisition campaigns for higher educated personnel from universities and technical schools.[38] DSM, for instance, promises *life-time employment* and high starting salaries for people who can belong to its core workforce.[39] Efforts of large firms to give higher-trained personnel the possibility to work from home fall in the same category. Some 12 per cent of big firms in the United Kingdom are already giving their executives this option.[40] A growing number of projects which are risky because of their high costs and the very specific knowhow needed are *subcontracted to university teams*. It gives the companies the possibility of avoiding employing too specialized a core workforce which cannot easily be adapted to changing company needs. This is especially true for very new areas like biotechnology, artificial intelligence or basic research in new materials.

Since new production concepts also imply new organizational experiments, the large firms increasingly search for an extension of their cooperation with schools of business administration and social faculties at universities.[41]

This subcontracting shifts the employment risk connected with experiments to specialized companies and universities, but at the same time increasingly ties these external actors to the companies.

For employees with a low qualification, the changing nature of large firms (see Chapter Three) is resulting in more flexible working contracts. The principle of *home-work*, as applied by some firms to executives, will certainly also be extended to the lower-skilled jobs. It is estimated that

[37] Interviews in France, June 1985.
[38] The large number of vacancies for higher-educated personnel with the large companies together with the often attractive terms may have the result that they 'skim off' the labour markets for highly trained professionals, to the disadvantage of the smaller firms which cannot offer the same conditions. See the research results of the Geographical Institute of the Rijksuniversiteit Utrecht and the Regional Economic Department of the Free University, Amsterdam.
[39] *Intermediair*, 6 September 1985.
[40] Some 60 per cent of the big firms expect that such forms of work organization will be possible within the second part of the 1980s. British study cited in *Social Europe*, no. 2/1984, p. 26, in which it is also noted: 'As to distance-work among salaried employees, only rather marginal experiments exist as yet.'
[41] See, for example, Siemens, *Annual Report, 1984*.

in the United States in 1985 5 million people already belonged to the category of workers having only a telecommunication link with the company.[42] DSM in Europe is explicitly looking for possibilities to employ home-workers. There is, of course, an enormous difference between the positions of low-skilled and high-skilled home-workers. The first will probably have a very loose formal link with the company, hardly any guarantees for continued work and probably few retraining facilities. With regard to high-skilled workers the firm's policy of making use of the technological possibility of home-work in certain core functions is aimed at binding these employees even more to the firm—a positive secondary working condition which distinguishes the firm from competitors.

Another element of flexibilization, making use of more *part-time workers*, seems to be a strategy predominantly confined to the lower-skilled jobs. SIEMENS expressed the intention to increase the share of part-time workers in the 1985–90 period from 5.5 per cent to 10 per cent. It is interesting to see that the 5.5 per cent share of part-timers with SIEMENS is well below the average share in most industrialized countries (Germany, Japan: 10 per cent; around 15 per cent in the UK and the USA and well over 20 per cent in countries like Sweden and Norway).[43] This indicates that manufacturing is 'catching up' with other sectors like the hotel industry or retailing and that the boundaries between the service sector and industry increasingly become blurred with regard to working conditions and qualifications.

Especially *women* are susceptible to rationalization or flexibilization of their jobs in the short run since they are the first to use modern data-processing equipment.[44] In this context there is no difference between women working in industry and those in services, although the women working in services by far outnumber those in industry. There seems to be a reversed relation between gender and qualifications with regard to the use of programmed equipment. Whereas higher-trained men use programmed equipment far more often than men with lower or no professional training qualifications, the opposite is the case with women: 10.2 per cent of the men who use equipment have had no professional training versus 16.9 per cent of women, whereas 36 per cent of the men using advanced equipment versus 20.8 per cent of the women are university or higher trained.[45]

To sum up, there seems a clear tendency towards *increased polarization* in qualification and work organization, especially in the larger firms in core technologies and with growing assembly-like operations. This polarization takes the shape of, on the one hand, a growing share of skilled personnel with permanent reeducation and, on the other hand, low-skilled

---

[42] See J. Verschure, *Intermediair*, September 1985.
[43] OECD Employment Outlook.
[44] IAB survey 1983, BIBB–AIB survey 1979, as cited in Salisch (1985), p. 18.
[45] IAB survey 1983, BIBB–AIB survey 1979, as cited in Salisch (1985), p. 21.

or even unskilled and flexible workers. The groups of semiskilled workers and middle management become increasingly obsolete. It has to be stressed that this certainly differs from company to company according to the internal balance of power, resulting in varying management strategies, and with considerable differences between sectors with regard to the pace of this process. The process takes a different shape, for instance, in the already highly automated chemical industry from the textiles industry.

However, automation of the production process with reprogrammable equipment as opposed to dedicated or 'hard' automation in general makes it possible to reduce the skill required for specific tasks even at higher levels in the organization.[46] Kern and Schumann[47] in this sense prefer to speak of a *segmentation* of the work organization as a new form of the already existing polarized structures. In this process of segmentation four societal groups can be identified: (a) those gaining from rationalization, (b) those having a job in core industrial sectors, but handicapped by age, sex or nationality, (c) workers in declining or increasingly periferal sectors and (d) the unemployed. The polarization between the first group (in Europe probably predominantly employed in the larger international firms) and the latter three groups is obvious.

With regard to the role of small and medium sized companies, the developments as analysed above severely affect their ability to remain or become an *independent* factor in new technologies. Small European firms have difficulties in obtaining higher-educated employees from universities due to the skimming off of the market by large companies. They are not able to invest in large in-house training facilities or provide their core workforce with attractive secondary conditions. This makes it difficult for them to deliver the training packages which are needed to sell high value added products. That relatively secure employment in core sectors can mainly be reached by proceeding along the narrow path of high basic education and permanent in-house, on-the-spot training within the companies[48] implies not only a discrimination against low-skilled, migrant and female workers, but also against small firms.

4.3. IMPACT ON REGIONAL INVESTMENT PATTERNS

The development of new core technologies will also have a far-reaching influence on the geographical location of production, both on a global scale

---

[46] See OTA (1984b), p. 111.
[47] Kern and Schumann (1984).
[48] Kern and Schumann (1984), p. 314.

and inside any given country. The restructuring actually taking place opens up new choices for the companies involved. When a new production line has been opened, it has often turned out to be more profitable to do this in totally new premises than at a location previously used for other purposes—not only because the investment costs for a new building are often lower (and the building time shorter) than for the renovation of the old one, but also because it would not be possible to adapt old premises perfectly to the exigencies of the new production technology. Where new facilities have to be built, new decisions have to be taken with regard to the location of the production. It is often expected that the new technologies will give rise to a more evenly spread economic activity than in the past, thus reversing the concentration on core industrial areas that has developed since the industrial revolution. There are a number of good reasons on which these expectations are based:

(a) The *minimum batch size* of profitable production will be considerably *reduced* by the application of new forms of flexible automation, as has been described by Chapter One. As a result, production in a number of decentralized plants would no longer lead to higher unit costs than centralized production. This opens up new avenues for the decentralization of production.
(b) New telecommunications equipment facilitates *control at a distance* and lowers communication costs. This would reduce the disadvantages connected with decentralized production.
(c) New machines increasingly become self-diagnosing and sometimes even self-repairing. When machines have to be repaired, it is often a whole module that is simply exchanged, which can be done by an unspecialized worker. This implies that less specialized services are needed in the neighbourhood and leaves the option to locate production outside an agglomeration area.

The development of new technologies thus tears down barriers to further decentralization. In principle, this increases the chance of Third World countries to attract additional investment, just as it may improve the odds of disadvantaged regions in the industrialized countries to become integrated into a new wave of industrialization.

The factors just mentioned are important. But do they give rise to a dominant trend towards decentralization? Or are there overriding countervailing forces that prevent the trend towards decentralization from manifesting itself? We believe that the latter is the case.

New technologies not only tear down the barriers to further decentralization, they also demolish the barriers to further centralization of production. Their social implications even demand further concentration. The following reasons have led us to this conclusion:

(a) Since traditional modes of production were more labour intensive, the demand for (a specific type of) labour often could not be satisfied by one production site (or would have increased the price of labour to an unacceptable extent). Consequently, production had to be spread over different sites in order to tap several local labour markets. Highly automated production will need much less labour (as we have seen) and will accordingly run less often into quantitative limits of labour supply.

(b) But the type of (specialized or especially versatile) labour required can more easily be found in agglomeration areas or attracted to agglomeration areas. Highly educated employees prefer to work in an environment with sufficient cultural attractions, a diversified education system for their children, and good traffic connections for national and international travel. These conditions only exist in (or near) agglomeration centres.

(c) R&D activities have become more and more important for the large companies. The creativity of scientists and technicians, however, depends very much on the culture in which they work. Face-to-face contacts with colleagues (from the same and other companies) have proved to be important even in a time of highly developed means of telecommunications. The concentration of microelectronics development in Silicon Valley in the United States is the most obvious example for this. High tech companies, therefore, will show more instead of less inclination to settle near other companies with similar production, because of the existence of specialized suppliers, education systems, services, and an environment of stimulating discussion. Since companies have realized that connections between production and research departments should be very close, it would not make sense to locate only the research departments in 'intellectual centres' and to place production somewhere else.

(d) New technologies are also used to reduce the negative environmental impact of industrial production. To the extent that this is the case, industries can become concentrated again that hitherto had to be spread in order to avoid pollution beyond a certain threshold.

(e) A still more important line of reasoning is the following: the application of new technologies has reduced the share of wages in total production cost (and promises to do so even more in the future). Thereby other cost items have become more decisive for competitiveness. The period of high interest rates in the early 1980s has made companies more aware of the high costs of inventories at all levels (raw materials, intermediate products, finished but unsold products), and recent management strategies consequently tackle this problem. New logistics that aim to keep inventories down are part and parcel of the

introduction of computer integrated manufacturing. Flexible manufacturing and on-line communication allow just-in-time production
and delivery of components at the moment they are needed for assembly.
To avoid delays, suppliers are expected to produce nearby.

Although new technologies make more decentralization possible, there are
strong arguments why the application of these same technologies favours
increasing centralization as well. Probably multinational companies can
combine the advantages of centralization and of decentralization in some
way. They may spread their worldwide activities over a number of
production sites in important regions — say Western Europe, North
America and South East Asia, eventually also Brazil. But *inside* these world
regions, they can be expected to concentrate research and production on
central locations. As a consequence, very few favourable effects can be
expected from the application of new technologies by multinational
companies in disadvantaged regions, either worldwide or inside the highly
industrialized countries.

The argument hitherto has remained at an abstract level. Let us therefore
add some empirical evidence that can illustrate the expectations formulated
above.

Information technology manufacturers in Europe are highly concentrated
in national metropolitan regions (e.g. Greater London and Greater Paris)
or major industrial agglomerations (Northern Italy, the regions around
Munich and in Baden-Württemberg in West Germany): 'In the United
Kingdom, approximately half of all establishments manufacturing
electronic computers and telecommunications equipment were located in
the South East planning region in 1979', and 'in France, Greater Paris
accounts for 80% of innovations in information technology industries,
and 60% of computer and component manufacturing employees'.[49]

With regard to office automation, the diffusion of the technology is
bound to spread more quickly in the bigger cities than elsewhere, because
the large offices are located in these areas. A study of the European
Community[50] concludes that for most European countries (explicitly
mentioned are France, Denmark, Germany and Belgium) companies
situated in the metropolis play a dominant role in office automation.

The new wave of automation has made it possible to shift component
production (hitherto often located in — or subcontracted to — low-wage
countries) back to the core industrial areas: THOMSON, for instance,
is ploughing back over 40 per cent of its revenues in chip production into
new manufacturing plants: 'much of the £120 million is being used to bring

---

[49] Gillespie *et al.* (1984), p. 86.
[50] DG-V (Directorat General for Employment, Social Affairs and Education of the
European Community), October 1984, p. 20.

chip assembly back to Europe. Automated plants at Aix-les-Bains and Maxéville will assemble chips more cheaply and reliably than subcontractors or subsidiaries in the Far East'.[51] In 1983 ASEA's chairman noted that the sales mix of the company 'changed considerably, away from developing countries to the industrialized nations and away from major projects to smaller ones with higher ASEA value-added content'.[52] This trend is not confined to the electronics industries. In the chemical industry, the shift from bulk production to specialty chemicals goes hand in hand with a concentration of production on core regions. BASF, for instance, is planning to concentrate its production more on West Germany and the United States, which is a reversal of its strategy in the 1970s when the company's objective was to produce more (of its bulk) chemicals in developing countries and ship them to Europe and the United States.[53]

The regional pattern of new activities of European multinationals is closely linked to other aspects of their social impact. A precondition for shifting production back to core industrial areas is often that the new automated equipment can be kept going 24 hours a day, 7 days a week, all the year round.[54]

A shift of production back to the industrial countries does not mean that employment there would be safeguarded. Production can come back just because it has become highly automated. There will accordingly be only a few additional jobs to handle increasing output, but very few jobs in the production process itself.

4.4. IMPACT ON WORKERS' PARTICIPATION AND TRADE UNION INFLUENCE

Advances in information technology have considerable impact on the structure of multinational enterprises (see Chapter Three) and on the

[51] *Electronic Times*, 30 April, 1985.

[52] ASEA *Annual Report, 1983*, p. 5.

[53] *Business Week*, 23 September, 1985, p. 35.

[54] A Philips officer, for instance, gave the following answer to the question whether automation could lead to a retransfer of production to the Netherlands: 'This can be the case. But then workers have to be prepared to work in three shifts, 24 hours a day and 7 days a week. Many factories in the Far East are already highly mechanized, but they work three times as long as factories here. It is therefore insufficient to ask: can we automate production? Extending working hours is part and parcel of this decision, because this is the next competitive weapon'. Henk Tolsma, Alles draait om flexibiliteit, in *Intermediair*, 1 June, 1984, p. 19.

structure of the labour force (see above). Both determine in how far workers and their unions can successfully influence company decisions that affect them directly or indirectly. Changes associated with the application of technology have direct consequences for trade union membership, the internal cohesion of the labour movement, the access of workers and shop stewards to information and the effectiveness of the unions' weapons in industrial disputes.

Impact on membership

The application of microelectronics will lead to a more qualified (albeit smaller) labour force (see section 4.2). Highly qualified (especially academically trained) labour is more difficult to organize than the traditional union constituency. These employees often come from an environment with little trade union tradition. They often show a relatively strong identification with the company. They apparently enjoy more freedom in the daily organization of their work than the rest of the workforce, and their activities often do not lend themselves to standardization or classification. It therefore becomes more difficult for trade unions to do anything for this layer of employees, who often see more chance in individual than in collective actions to improve their position. This perception is justified to a certain degree because scarce specialists enjoy a rather strong bargaining position in the labour market, especially in the early phase of the development of core technologies when the offer does not match demand.

A shift towards core technology activities and the concomitant shift in the qualification of the labour force tends to lead to a decline in trade union membership and strength. Other factors connected to the application of new technologies work into the same direction. The increasing use of *home-work*, for instance, made possible by the use of new means of telecommunication, leads to the incorporation of individuals into the labour force who are very difficult to organize. The smaller average size of factories creates additional difficulties, because workers in small factories are more difficult to organize than workers in large factories (where trade unions employ full-time shop stewards).

Internal cohesion

To the degree that the better educated employees do become organized, they mostly join white-collar unions which often are in conflict with other unions. Conflicts between and inside the unions provoked by the application of new technologies are very common. Companies that introduce computer aided equipment have to make a choice whether the machinery will be programmed on the floor by skilled labour or by the people in the programming division.[55] In the UK this means a choice between different categories of workers, organized in different unions, with resulting conflicts between these unions. The most violent conflicts during strikes at TALBOT (PSA) car factories in France in the early 1980s were not between workers and management, but between different groups of workers which were differently affected by the introduction of new technologies.

Large-scale rationalization also puts the workers of different plants against each other. In spite of possible negative consequences to employment and working conditions, workers often have a strong interest in the diffusion of new technology to their factory. If not modernized, their factory risks obsolescence and becomes an early candidate for closure because of lack of competitiveness.[56]

Access to information[57]

The integration of computer technology with telecommunications made it possible to separate, geographically, closely related work processes and to have parallel work done by workers who are not in direct contact with each other. Modern information technology increases in asymmetry between the information available to management on the one hand and to shop stewards and employees on the other. Whereas computerized *management information systems* increase the amount of information on which management can draw, shop stewards get *less* information. They depend very much on face-to-face communication on the work floor. The more the work process becomes divided into geographically separated parts, the more difficult it becomes for them to get an idea of the whole process and to evaluate management proposals accordingly.

[55] See Shaiken (1984).

[56] This mechanism was part of the background of a 1985 strike at Vauxhall in the UK, where workers demanded further investment in the plant to increase production and opposed the company's decision to import cars from other European sites of Ford whenever domestic demand exceeds domestic production. *Financial Times,* 10 September, 1985.

[57] See Levie and Moore (1984) on several 'best practice' examples throughout Europe.

Effectiveness of trade union actions

While a declining membership, diminishing cohesion of trade unions and reduced access to information tend to undermine trade union influence, there is one important counter tendency. The logistics of the production process have become streamlined to such an extent that it has become very vulnerable to (trade union) action. Increasing specialization has often led to a situation in which one single supplier produces crucial components for a number of companies. Trade union action in such a unit could effectively paralyse production of a whole industry. This seems to give a strong bargaining weapon to the trade unions. But it is a big question whether they will keep it. In several European countries, legislation is under way (or under discussion) on how such an 'unfair' influence can be curbed.

The consequences of the introduction of new technologies for the overwhelming majority of trade unions thus will be rather negative. Not only will the total labour force be reduced, its composition will also change in such a way that the degree of organization probably will decline. Their internal cohesion will diminish, and access to much important information will be denied to them. Where their influence tends to increase as a consequence of increased vulnerability of the production process, legal steps are taken to prevent such an increase happening.

4.5. CONCLUSION

We could only deal with some aspects of the social impact of the move of European multinationals to concentrate more on core technologies. Unless technological development leads to a very large number of new products and stimulates a new wave of fast economic growth, the employment impact of new technologies will be negative for most European countries. Most of the remaining jobs, however, will have to be more versatile and will generally demand a higher education than the average work hitherto. Working conditions will change so that employees will have to accept higher flexibility, and some of the accomplishments of the labour movement in the past, such as a reduction of shift and night work, will be abolished. This is the precondition for a retransfer of component production to the industrial centres, which actually can be witnessed—not so much in the form that plants are closing in developing countries and the production is taken over by factories in Europe or the United States, but in the form that the production of new generations of products is planned to take place in the industrial countries, based on new process technology (and labour relations) that make these locations competitive again. Unfortunately, there has not been enough research on what such a shift in the long run will mean for the ability of developing countries to earn foreign exchange, for their long-run economic development prospects and for worldwide further economic growth and stability.[58]

---

[58] Actually, there is some research under way by the ILO on this topic.

Chapter Five

International Competitiveness of European High-Tech Firms

The chapter on the social impact of new technologies has shown that the developments described have a rather ambivalent impact on employment. Much employment is lost because of the rationalization and automation of production and services. Some new employment is created in companies that produce the new equipment. How the new and the remaining employment will be distributed internationally, is mainly a question of the international competitiveness between companies in different regions.

This chapter will analyse the international competitiveness of European hightech multinationals from three angles. First, some general remarks will be made that are applicable to all sectors, more specifically with the problems of the alleged smaller size of European multinationals, the higher degree of diversification, and the technological or market oriented determinants of competition. Second, the competitiveness of European companies in four of the 'core technologies' will be analysed more in detail. Third, the general social setting that is of crucial importance for international competitiveness will be dealt with. The university-industry, the cost of capital, the quality of management and the general problem of human resources are considered.

This chapter will exclusively deal with competition between American, Japanese and European companies. Since the application of new technologies will concentrate economic activity in highly developed countries again,[1] it is the competitive position of European firms in comparison with American and Japanese firms that matters. Competition from the so-called 'newly industrializing countries' is only felt in a few sectors, and mostly comes from American, Japanese or European companies in these countries, not from indigenous firms. Competition from socialist countries will become even less relevant than in the past, because these countries have considerable difficulties in applying core technologies on a large scale.

---

[1] See Ohmae (1985).

5.1. SPECIAL FEATURES OF EUROPEAN MULTINATIONALS

5.1.1. Smaller size: a problem of the past

During the 1960s, a discussion on the international competitiveness of European companies took place that was to some extent similar to the present one. At that time as well, the 'American challenge' was discussed,[2] and extensive analyses of European competitiveness were made.[3] One of the results of that debate was the observation that European companies were not big enough in comparison with their American competitors. Of the first 100 of the international *Fortune* list, only 29 were European companies in 1965 (against 69 American companies). In 1980, however, 42 were European against 45 American companies. Leaving the oil companies aside (20 of the 45 American companies), there were more European than American companies in the list of the world's largest companies.[4]

Accordingly, size is no longer the major problem. But to gain size, most European companies had to choose a different strategy from that of their American counterparts. Given the relatively small national home markets and stiffer competition abroad, many European companies tended to diversify and become active in a broad range of activities. As a result, European companies, especially in the electronics and the chemical industries, are more diversified than American companies. This has obvious consequences for their competitive position.

5.1.2. Consequences of higher degree of diversification

Since European companies are active in a large variety of fields, they have to distribute their R&D expenditure over many areas, with the result that in any individual field companies will spend less on R&D than their more specialized competitors. Although European companies do not spend less on R&D as a proportion of total sales (see Table 5.1), their R&D investment is more thinly spread. For this reason, they will less often reach the 'critical mass' of investment that is necessary to achieve a breakthrough in a specific

---

[2] Servan-Schreiber (1967).

[3] See Uri (1971).

[4] Compiled on the basis of the two rankings of the largest American companies and of the largest companies outside the United States; *Fortune*, July and August 1966, May and August 1981.

Table 5.1   R&D as a percentage of total sales with major European, US and Japanese multinationals, 1980–1984/85

	1980	1984/5
10 European firms		
SHELL	0.6	0.7
ICI	3.7	3.2
HOECHST	4.3	4.9
PHILIPS	6.9	6.7
SIEMENS	9.0	8.8
VOLKSWAGEN	3.7	3.2
NIXDORF	7.7	9.8
BULL	9.9	9.9
KRUPP	1.1	1.3
CIBA–GEIGY	7.9	9.2
10 American firms		
EXXON	0.5	0.8
DOW CHEMICAL	2.4	4.7
DU PONT	3.5	3.9
GENERAL ELECTRIC	2.7	3.8
WESTINGHOUSE	2.2	2.3
GENERAL MOTORS	3.9	3.8
IBM	6.6	6.9
DIGITAL EQUIPMENT	7.9	10.7
BETHLEHEM STEEL	0.7	0.8
ELI LILLY	7.8	11.3
8 Japanese firms		
MITSUBISHI CHEM.	3.6[‡]	2.9
KOMATSU	2.8[†]	4.2
HITACHI	3.8	5.3
MATSUSHITA ELECTRIC	3.5	4.4*
NISSAN	2.6	4.0
NEC	5.0	12.7
SONY	5.3	7.8
SUMITOMO ELECTRIC	2.0	3.2

*1983; [†]1981; [‡]1982.

field. A large spread of activities implies a *followers' strategy*, a strategy that aims at following the technological leader at some distance, taking care that the distance does not become too great. Although such a strategy will seldom allow the companies concerned ever to become the first in their sector, it can be very rational. The most risky developments will be carried out and paid for by others. Companies will only have to take up what has turned out to be viable. It is estimated that the costs for an 'imitator' to follow technological developments elsewhere are only 40 per cent of the costs of the forerunner. As long as companies have a tight grip of their own national or regional market, being 'a good second' will not have negative consequences for their international competitive position. The fact that they can allow themselves to spend less on R&D (in any individual field) may sometimes even give them a price advantage.

This argument has been valid for some time, but it is an open question whether it still holds true, for at least three reasons. It was true as long as production technology was more or less standardized and stable, which was the case during much of the 1960s. New technological developments actually, however, affect first of all *production* technology. If a company is able to make use of a flexible production technology that allows it to produce a new product at an acceptable price much in advance of others, it can conquer the market (if not protected) to the detriment of the late-comers. A second reason is that international competition has become much stronger since tariff protection of individual markets has been reduced and international marketing organizations of multinational companies (as well as the network of international cooperation agreements) have expanded. As a result, a company cannot wait any longer to apply new technologies, because it cannot be sure that the market will wait until it comes up with a comparable product.

Increased international competition and stagnating demand in many areas have forced companies to make use of any possible outlet to increase sales. These efforts have shortened the life-cycle of products considerably. This implies that it has become more difficult for imitating companies to catch up with the forerunner. When product life-cycles were longer, the 'follower' would, after a while, be able to offer a similar and equivalent (and sometimes even improved) product to that of the leading company. With shorter life-cycles, however, the leading company will have a *new* product ready at the moment that the follower is able to market the former generation of products. In this way, a temporary setback turns into a permanent technological gap that sooner or later also has an impact on the public image of the company concerned. As long as it could offer reliable equipment, free of the flaws that the first types of a new product often show, it could convey the image of a solid, advanced company, though not a forerunner. If it starts to lag behind continuously, however, it will get the image of a comparatively backward producer.

The followers' strategy has become much more risky, as a consequence, than it used to be. But the higher diversification of European companies is not only a disadvantage in the race towards innovation. The innovation process consists not only of an original invention and the consequent development to realize a new product (or process)—the other necessary step is the *diffusion* of the invention. We have stressed the fact that the *core technologies* are applicable in a wide range of products and processes. What European companies may be much better able to do than American companies is to use advances in one specific sector in other sectors as well. A European company that may have chosen a followers' strategy (or cannot afford another) may nevertheless be the first to apply the innovation in *related* areas, if it has organized its internal communication in such a way that this kind of technology transfer is encouraged. The potential for this kind of technology transfer has considerably increased with the blurring of technological boundaries between different sectors. Few European companies, however, have institutionalized routines to use advantages resulting from their more diversified structures in a more systematic way.

5.1.3. Determinants of competition:
market dominance versus technology dominance

To arrive at a conclusion with regard to the international competitiveness of companies or economies, some notion of the different weight of different areas of activity is needed. Not all sectors and all product groups are equally important. Some fields are more important for a company's or an economy's future than others.

J. Malsot argues that the crucial steps in a chain of production vary depending on the phase of the product cycle.[5] In the early phase of the life-cycle of a product (technology, industry), companies that dominate the forefront of technological development dominate the whole chain of production. In a later phase, however, when the project has become mature and does not undergo profound changes any longer, the more consumer oriented last phases of the production chain become dominant, and companies that dominate commercialization are able to dominate the whole chain of production. During the first phase of the life-cycle, the technological performance characteristics tend to dominate the clients' purchasing behaviour. As the technology becomes more and more diffused, other aspects such as service, marketing channels and publicity become more important.

The shorter product cycles become, the more difficult, however, will it be to differentiate between these different phases. They will tend to overlap more and more. A company that wants to bring a new product onto the market will have to do this immediately in *all* the major markets. This implies that marketing power, from the first phase on, will be important and can be translated into a technological advantage. Large sales produce enough cash to invest in research and development which

[5] Malsot (1980), p. 33.

may lead to a technological edge. Or market access can be exchanged for access to the most advanced technology of a cooperation partner. Many cooperation agreements between European multinationals and partners from overseas follow this pattern.

Since the market environment has a significant impact on the international competitiveness of an industry, some market data on high technology markets in Europe will be given and will be compared with those of the United States and Japan. The dynamics of demand in home markets have an important conditioning effect that has a strong impact on the international competitiveness of the companies concerned. In a home market with a large diffusion of sophisticated products, new products normally will be accepted very fast, and a buoyant demand can be expected. In a market with little demand, companies can hardly prosper.

5.2. COMPETITIVENESS IN FOUR CORE TECHNOLOGIES

The international competitiveness of European multinationals in four areas will be described: in semiconductors, in telecommunications, in the robot industry and in biotechnology. This chapter tries to depict the competitiveness of European multinationals as a whole, without much differentiation according to companies and countries.[6] The large differences between the individual companies (and the countries in which they have their origin) are described in the next chapter.

5.2.1. Competitiveness in semiconductors

The West European semiconductor industry lags so far behind American and Japanese producers that it is hardly mentioned in international studies of the competitive strength of major producers.[7] This has not always

---

[6] The differences with regard to the innovation potential probably are larger between one European country and another, though, than they are between the United States and Japan on the one hand and Western Europe as a whole on the other. In a detailed study of determinants of innovation, Henry Ergas comes to the conclusion that the situation in West Germany, Sweden and Switzerland does not differ significantly from the situation in the USA or Japan, whereas it is much less favourable in France and Great Britain. See Ergas (1984a), p. 32.

[7] See 'International competition in advanced industrial sectors: trade and development in the semiconductor industry', a study prepared for the use of the Joint Economic Committee, Congress of the United States, Washington, 1982; Semiconductor Industry Association (1983); US Department of Commerce (1983b).

been the case. During the 1950s, the European semiconductor industry occupied 'a position equal to that of the American industry and superior to the Japanese industry. . . . Later in the 1960s, however, the European industry declined and lagged more and more behind the American industry. The situation did not improve during the 1970s and early 1980s.'[8]

The interaction of three major factors has been identified as the major reason for this relative backwardness: size and sophistication of demand, the structure of the industry, and the timing, type and extent of government support. During the 1930s and 1940s, European firms were among the first producers of new semiconductor devices.[9] During the 1950s, Europe remained mainly self-sufficient in semiconductors. It was the relative success of European companies in this early period that contributed to their failure in the 1960s when integrated circuit technology was introduced. As relatively successful companies, the European companies were good at managing incremental technological change, but had difficulties in mastering radically different technologies. They produced first of all for the consumer goods industry, while American industry faced increasing demand from the military and from the computer industry. To satisfy the demand for more sophisticated devices, a number of new companies were formed in the United States which could survive because of steady government support to R&D and a guaranteed market. These new companies were more committed to the new technology than the existing companies (in Europe as well as the United States). This constellation of (a) a more sophisticated demand, (b) the creation of new firms committed to the new generation of technologies, (c) strongly supported by government, has given the American semiconductor industry an advantage from which its strong position in the 1960s and 1970s derives, even after the role of military purchases as a per cent of total sales became less than 10 per cent.

The per capita semiconductor consumption (measured in dollars) of the European countries in 1985 was considerably lower than in either Japan or the USA: on average in Europe per capita consumption was $14,[10] whereas in the USA $53 worth of chips were consumed and in Japan even $70. In the 1978–85 period the consumption of semiconductors grew more than threefold in the USA and Japan but only twofold in Europe.[11] As a result the share of European semiconductor consumption has fallen from around 30 per cent of world consumption in 1974 to around 16 per cent in 1983. And this is even more dramatic in the more sophisticated parts

---

[8] Malerba (1985), p. 3.
[9] Malerba (1985), p. 3.
[10] $21 in Germany and the UK, but $13 and $12 in France and Italy respectively.
[11] *Financial Times*, 24 June 1985.

Table 5.2   Market share of USA, Japan and Western Europe (%), 1982 (1983)

	USA (1983)	Japan (1983)	Europe (1983)
Semiconductors	56.6 (56.4)	26.3 (27.3)	17.1 (16.2)
— Integrated circuits	62.8 (61.8)	22.1 (24.3)	15.0 (14.0)
— Memories	71.7 (72.3)	16.3 (16.9)	12.1 (10.8)
— Microprocessors	73.1 (63.8)	19.2 (27.3)	7.7 (7.2)

Source: Based on *Electronics*, 12 January 1984.

Table 5.3   World market share and geographical distribution of markets of major world merchant semiconductor producers, 1978 (%)

Company	World market share	Geographical breakdown of sales USA	Japan	Europe	Other
Texas Instruments	11	55	10	31	4
Motorola	8	62	5	25	8
PHILIPS (inc. Sign.)	7	24	4	63	9
NEC	7	8	77	4	11
Hitachi	5	6	80	2	12
National Semiconductors	5	65	5	19	11
Toshiba	5	6	70	4	20
Fairchild	5	63	3	18	15
Intel	4	59	3	27	11
SIEMENS	3	12	0	78	10
Others	40	—	—	—	—

Source: Nomura Research Institute, 1980; cited in Dosi (1981), p. 65, and OTA (1983), p. 143.

of the semiconductor market. Whereas about 16 per cent of all semiconductors are consumed in Europe, only 7 per cent of microprocessors are sold there (Table 5.2).

The European market is not only smaller, it is also growing more slowly than the American and the Japanese markets, which creates a comparatively unfavourable environment for European semiconductor producers, whose sales concentrate heavily on regional home markets, as Table 5.3 indicates. The comparatively large share of their total production that American semiconductor producers sell in Europe is a reflection of the weak position of European producers. During the 1960s and 1970s, American chip producers invaded Western Europe. The fact that Japan did not allow foreign companies to penetrate the domestic market to such a degree is regarded as one of the major reasons why the Japanese electronics industry was able to catch up during the 1970s and early 1980s, whereas the European industry was not.[12]

[12] Malerba (1985).

Table 5.4   Export orientation and import penetration in integrated circuits, 1982

	Exports as % of production	Imports as % of consumption
USA	19.1	7.2
Japan	28.7	13.9
Europe	17.7	65.8

Source: OECD (1985c).

Table 5.5   Semiconductor production, estimates and projections (billions of dollars)

Production owned by	1979	1981	1983	1984	1985	1988	1992	Annual average growth (%) 1979–83	1984–92
US	9.8	14.2	17.2	23.1	26.7	38.6	70.5	15.1	15.0
%	68.5	68.6	65.1	63.1	63.9	63.7	59.1		
Japan	2.8	4.9	6.7	9.2	11.2	18.2	37.2	24.9	19.1
%	19.2	23.7	25.5	25.9	26.8	30.0	31.2		
Europe	1.7	1.5	2.1	2.6	3.0	4.0	6.6	5.2	12.3
%	12.0	7.2	8.0	7.3	7.2	6.6	5.5		
Other	0.04	0.1	0.3	0.6	0.9	1.8	5.0	68.6	30.3
Total	14.3	20.7	26.3	35.5	41.8	60.6	119.3	16.5	16.4

Source: US Department of Commerce, International Trade Administration (1985).

While the import penetration of Japan declined from more than 20 per cent in the 1970s to around 10 per cent, import penetration of Europe increased from 40–50 per cent during the 1970s to more than 65 per cent (see Table 5.4).

Japan increased her world market share at the expense of European producers. In Europe, however, the rise of Japan's share of sales (from 3 to 10 per cent in the period 1977–83) was mainly at the expense of American suppliers, whose market share declined from more than 60 per cent in 1977 to around 50 per cent in 1983. The market share of European firms remained fairly stable at a level between 35 and 40 per cent. The world market share of semiconductor production owned by European producers, however, is expected to decline further, as Table 5.5 shows.

5.2.2. Competitiveness in telecommunications

One of the three mass markets of semiconductors (besides consumer electronics and the computer industry) is the telecommunications sector. Only in Europe (compared to Japan and the United States) does

Table 5.6   Percentage share of the EC in total OECD trade in telecommunications equipment,* 1978–84 ($ million at current prices)

	1978	1980	1984
Exports			
EC	47.1	46.8	32.8
USA	17.3	16.2	17.5
Japan	23.3	23.0	34.0
Imports			
EC	47.9	49.0	30.7
USA	30.2	29.1	50.0
Japan	3.5	2.6	2.9

*Category 764 of the Standard International Trade Classification (Rev. 2).

telecommunications as an end-use market for semiconductors surpass computer and consumer electronics. This underlines the importance of the telecommunications industry as a demand-pull factor for semiconductor technology in Europe. In telecommunications, European competitiveness has been judged to be much better than in the case of semiconductors itself. In world markets for telecommunications systems the European exporters indeed had a strong position until the beginning of the 1980s as can be witnessed from Table 5.6.

The data in Table 5.6 show, however, that the position of the EC countries declined rapidly in the first half of the 1980s, with Japan taking a larger share of world export than the EC countries together (even considering intra-EC trade!) in 1984. The growth of Japanese exports and Japan's positive trade balance have almost exclusively come from the increased penetration of the American market, whereas the European countries only have a positive trade balance with developing countries. The latter is largely the legacy of the colonial past, when European colonies had little choice but to order their telecommunications equipment from companies of their respective colonial motherlands. After independence, it would have been rather costly to change the main suppliers. With the change from analogue to digital switching, however, competition has become more open. But even in 1984 the (declining) positive trade balance of Europe with developing countries still more than compensated for the negative balance with regard to the United States and Japan.

A continuation of the strong position of the EC producers in Third World markets is not enough, though, to lessen the downward trend in Europe's competitiveness in telecommunications. The US and, especially, Japanese producers seem increasingly able to penetrate each other's markets which together account for more than 60 per cent of the world market. European firms have not been very successful yet in these markets. This negative trend can have important spillover effects to other sectors of the economy.

Investment by European countries in R&D is no less than that by the United States and Japan. On the contrary, it is estimated that European countries spend twice as much on R&D in telecommunications as the USA or Japan.[13] The development of a full range of digital switching systems by the nine competing European firms necessitated a joint investment of $6–7 billion, whereas the four major competing American firms (AT&T, GTE, ITT and Northern Telecom) together spent only $3–4 billion. Japan has the smallest number of competing suppliers (NEC and Fujitsu) with total investments of about $1.6 million, enough to achieve comparable objectives.[14] Technologically, European telecommunications companies do not lag behind. They do participate with some success in the competition to become a second or third supplier of the seven regional 'Bell sisters' that were created after the splitting up of AT&T.[15] Whereas the market for switching equipment in Europe is a political market in which the 'national champions' are protected against too stiff competition from the outside, the procurement of the 'Bell sisters' seems to be a more reliable indicator of the price/performance relationship of competing companies. The fact that European telecommunications companies dispose of highly advanced technology is also underlined by the fact that in 1985 THOMSON (in a consortium with GTE Corporation) got the $4.3 billion order for the new mobile battlefield telephone system for the US Army, the largest order ever given by the Pentagon to any European company.[16] It is also significant that the most important competitor in this case was another European company (PLESSEY, in cooperation with Rockwell International Corporation).

While a number of European companies succeeded in keeping a competitive edge in telecommunications technology, they appear handicapped by the small size of their respective national markets, the different standards of European telecommunications services and the lack of sophistication of European markets. The lack of standardization does not seem to be such an obstacle that it discriminates against European companies. When ITT announced that it would no longer compete for American orders for its System-12 exchange (before it sold the majority share in its telecommunications business altogether to CGE), it gave as

---

[13] Yankee Group (1983), p. 151.
[14] Arthur D. Little estimates, 1983.
[15] Examples are Siemens, Ericsson and even Plessey.
[16] *Fortune* 5 August 1985, p. 38.

Table 5.7   Introduction of 'integrated services digital networks'

	Experimental installations	Limited services through pilot projects	Official intro- duction of new services
Belgium	1984/5	1988	1989/90
Denmark	—	—	—
France	1986	—	1988
Germany	—	1986	1988
Ireland	1986/7	1988	No decision
Italy	1984	1987/8	1990
Netherlands	1987	After 1988 theoretically possible but no decision taken yet	
Spain	1985	1987	1988
Sweden	1984	1987/8	Depending on demand
Switzerland	—	1987/9	No decision
UK	1983	1984/5	Pilot projects are regarded as regular services
USA	—	1986/7	1986/7

Source: Deutsche Bundespost, Mittelfristiges Programm für den Ausbau der technischen Kommunikationssysteme, Frankfurt, 1986.

one of the reasons that it was too costly to adapt the design to the different specifications of the different regional telephone companies in the United States![17] But the lack of sophistication of the European market is a serious handicap. The use of the telephone system for other purposes than traditional voice communication is much more advanced in the USA and in Japan.

The very fact that European countries protected their national companies often led to a situation in which new services were only introduced when domestic companies were able to provide the corresponding equipment — thereby reducing the incentive to speed up the development for new equipment. Many European companies (like SIEMENS, and like ERICSSON with its mobile car telephone) sold their most advanced equipment (e.g. digital PABX exchanges) first in the American market. European policy makers have recently realized the negative interplay between protection and competitiveness with the result that there are strong pressures in favour of more deregulation (see section 6.3).

To create a stronger challenge for European telecommunications companies, the national PTT organizations as well as the European Community (see the RACE programme, Chapter Seven) are hastening to develop Integrated Services Digital Networks (ISDN) which will allow the

[17] *NRC-Handelsblad* (Rotterdam), 19 February 1986.

communication of voice, moving images, text and data through the same channels. Table 5.7 surveys the plans in different countries.

The table shows that most European countries do not lag much behind the USA with regard to the date of introduction of new services. The plans might look good in Europe, but the level of investment per capita in telecommunications equipment has been considerably lower in the EC than in either Japan or the USA. In 1981 total telecommunications investments in Europe (measured in dollars) were 70 per cent of the Japanese per capita investment and only 40 per cent of US investment.[18] Plans for the new ISDN services, therefore, have to be accompanied by a considerable rise in overall spending in order to have the same effect in Europe as in either Japan or the USA and to effectively use the telecommunications sector as a lever to boost the ailing European information sector as a whole.[19]

Only at a high price has Europe been able to retain its relative competitiveness in telecommunications. European producers incurred high R&D costs for the development of digital switching equipment, while fragmented markets hardly allowed them to earn this investment back. This implied that the money had to come from somewhere else. To some degree, the burden was taken over by the national PTT and, indirectly, the service consumers who were charged high rates. It has been estimated that telecommunications equipment in the early 1980s cost the West European authorities 60 to 100 per cent more than their American counterparts.[20] The comparatively high price of telecommunications services in Europe, however, has a negative impact on the competitiveness of European industry as a whole and slows down the expansion of the network and the demand for new services. In this way telecommunications companies ultimately hurt themselves, because their markets do not expand fast enough.

In other cases, the R&D expenditures were carried by the companies themselves with the immediate result that they had to spend less on R&D in other fields. Since the telecommunications equipment producers in Europe are by and large identical with the European chip producers, there may even be a causal link between the relatively good performance in telecommunications and the bad performance in semiconductors. Since European telecommunications markets were safe, European companies may have spent large amounts of R&D in a field where sales were guaranteed (and public services shared the financial burden). But at the same time less money could be spent on non-customized chip development with the results described above.

---

[18] Arthur D. Little, cited in *Financial Times*, 6 January 1986.
[19] See Mackintosh (1986).
[20] See *The Economist*, Telecommunication Survey, 23 November 1985, p. 28.

A fundamental change will only come about when the number of European telecommunications companies is finally reduced. Such a *shakeout* is actually taking place.

5.2.3. Manufacturing technology

To illustrate Europe's role in manufacturing technology, the use and production of robots will be taken as an indicator. However, robots are only one component of computer aided manufacturing. A comparison during the first half of the 1980s of the number of robots installed was a little fanciful because so-called 'island automation' dominated the efforts to introduce flexible automation into the production process, i.e. the job of one or two workers was taken over by a reprogrammable machine, but this was not necessarily linked to other machines. More and more, however, these machines have become linked to each other. In such integrated systems of automation, computer aided design plays a key role, because it is here that the common database is created that will serve all the three sectors.

Whereas the use of robots was considered crucial for increasing productivity in the early 1980s and the international 'robot race' was emphasized accordingly, the emphasis has shifted somewhat towards the design stage. Here the European position is weaker than in robot

Table 5.8   Number of robots installed

	USA	Japan	Germany	France	UK	Italy	Sweden
1981[a]	4 500	9 500	2 300	790	713	450	1 700[b]
1982[a]	6 250	13 000	3 500	950	1 152	790	1 300[b]
1983[c]	8 000[d]	16 500[d]	4 800	1 920	1 750	1 800	1 850
1984[c]	13 000	21 000[e]	6 600	2 750[f]	2 255[f]	2 600[f]	2 400
Annual growth rate 1981–4	72%	55%	47%	62%	54%	119%	103%

Sources:
[a] OECD (1983b), p. 50.
[b] *OECD Observer*, no. 123, July 1983, p. 11: data revised downward as a result of definitional changes.
[c] AFRI (French Robotics Association), Statistiques 1984, p. 4.
[d] British Robot Association, *Robot Facts*, December 1983 (Fig. 1).
[e] Estimation of the Associations Européennes de Robotique, cited in AFRI, Statistiques 1984, p. 4.
[f] The British Robot Association, December 1984 (as cited in *Industrial Robot*, March 1985, p. 30) provides the following figures for France, the UK and Italy respectively: 3 380, 2 623 and 2 700.

Table 5.9   Number of robots in use per 10 000 workers in industry

	USA	Japan	Germany	France	UK	Italy	Sweden
1980[a]	3.1	8.3	2.3	1.1	0.6	n.a.	18.7
1982[b]	5.0	13.0	4.6	1.9	1.2	1.0	29.9
1984[c]	4.3	32.0	5.7	4.3	4.8	3.5	n.a.

Sources:
[a] Belgian Institute for Regulation and Automation, Metalworking Production, April 1983, pp. 80–4 (based on employment data of the OECD).
[b] *OECD Observer*, no. 123, July 1983, p. 13. Different sources show different figures: The journal *Industrial Robot*, **11**, no. 1 (March 1984), cited in Economic Commission for Europe (1985), Table 35, p. 46, provides figures which are much lower for all countries except for the figures on the number of robots, presented in Table 5.8 (with the exception of the figure for Italy: whereas 790 robots were installed according to the OECD, the ECE mentions only 700 robots).
[c] British Robot Association, December 1984, cited in *Industrial Robot*, March 1985, p. 30.

technology. And this weakness hampers the advance of automation in *all* sectors of any company.

For the future of the European equipment producers, it can be a major problem that the largest producers of robots are the largest users at the same time. Their machine-building (automobile) background could turn out to be a disadvantage in the future, as they have to face the challenge of the integration of different technologies, with electronic components becoming more and more dominant for the whole system.

Whatever the definition of robots that is used, it is obvious that Japan has a clearcut advantage with regard to the numbers of robots installed. The United States, however, does not have a comparably prominent position. The four largest European countries (Germany, France, the UK and Italy) together show a larger number of robots installed than the American industry. (See Table 5.8.)

If we look at *robot density*, the number of robots in use per 10 000 workers in industry, we find that European countries by and large do not lag behind the United States. The robot density is even higher in Sweden and in Germany and more or less the same in France, the UK and Italy. (See Table 5.9.)

Europe is not at a disadvantage with regard to the number of robots, either absolutely or in comparison to the number of workers in industry. What can be a problem, however, is the strong concentration of the robot population on the motor industry. Especially in the case of Germany, the country with the largest number of robots in Europe, the concentration on the transport and metal products industries is very high (with 62 per cent of all robots in 1981).[21]

---

[21] Commission of the European Communities (1985d), p. 8.

Table 5.10   The use of robots in the motor industry

Manufacturer	No. of vehicles produced (1982)	No. of robots (1982)	No. of robots per 100 000 vehicles
MERCEDES (D)	480 000	550	114
BMW (D)	360 000	350	97
VOLKSWAGEN–AUDI (D)	1 538 000	950	62
FIAT (I)	1 186 000	640	54
GENERAL MOTORS (USA)	4 630 000	2 300	50
TOYOTA (J)	3 145 000	1 400	44
NISSAN (J)	2 408 000	1 000	41
MAZDA (J)	1 110 000	430	39
CHRYSLER (USA)	967 000	360	37
FORD (Europe)	1 233 000	400	32
RENAULT (F)	1 719 000	450	26
PEUGEOT (F)	1 423 000	350	25
FORD (USA)	2 192 000	500	23

Source: Peugeot SA (cited by the Commission of the European Communities (1985d), p. 9).

While West German car manufacturers, as a consequence, have an outstanding position with regard to the robotization of their production process (as Table 5.10 shows), other industries lag far behind. Since 1982, the American car manufacturers probably have caught up. Still, European manufacturers are in a good position regarding the number of robots per 100 000 vehicles produced. It may be due to this competitive pressure that Ford Europe has a larger number of robots per vehicle produced than Ford USA. Just as European car manufacturers have been travelling to Japanese plants to learn about the latest progress in factory automation, Japanese engineers now visit the famous 'Hall 54' of VOLKSWAGEN at Wolfsburg to get acquainted with the advances made there.

This strong position of European car manufacturers, however, is somewhat ambiguous. With the exception of ASEA, the biggest robot users in Europe are at the same time the largest robot producers: VOLKSWAGEN in Germany, RENAULT's subsidiary ACMA in France and the FIAT subsidiary COMAU in Italy. For the future, this may have a four-fold negative impact on innovation in Europe:

(a) The OECD has warned that such a domination of the market by the largest users may *slow down the spread of robot technology* by not encouraging applications in other firms.[22]
(b) It will not only slow down the diffusion of robots, but also their development. The car companies will probably not develop or market *general* advanced robots for assembly purposes. But at the same time they do discourage others to do so, because the hitherto largest segment

[22] OECD (1983b), p. 29.

of the market (the automobile industry) is already occupied by their captive producers.

(c) The fact that the largest users produce mainly for their own use is an impediment to the standardization of equipment. Manufacturers producing mainly for their own use have relatively little interest in standardization. The resulting lack of common standards, however, is a barrier to further innovation.

(d) The higher sophistication of future generations of robots does not depend so much on advances in mechanical engineering as on advances in sensor technology and software. The machine-building and metal-working background of the large European producers is of little use in this field. Since European electronics companies have not developed a strong foothold in robot production, European robot producers may be forced to turn to American and Japanese electronics companies to get the electric 'brains' for their robots.[23] In the long run, this may make them mere 'bridgeheads' of American and Japanese companies competing in the European market.

The dominance of European car manufacturers as robot producers can also make it more difficult in Europe to link the robots to integrated CAD/CAM systems. If this is true, European manufacturers would face more difficulties in overcoming 'island automation' and introducing 'intrasectoral' and 'intersectoral' automation with the considerable productivity gains involved.

While computer aided *manufacturing* is reasonably developed in Europe (at least in comparison with the United States), European firms are unquestionably lagging far behind the United States and Japan in the use of CAD. Not only is the level of CAD utilization much lower in Europe than it is in the USA, but the CAD equipment that is used in Europe is mostly bought from American companies, as Table 2.5 has shown. This, again, underlines the comparative strength of Europe in 'traditional' areas and the weakness in the electronics sector.

5.2.4. Competitive position in biotechnology

Whereas the other technologies discussed in this chapter all were applications of microelectronics, biotechnology is a different set of

---

[23] Siemens, however, claims that 'major robot manufacturers throughout the world use Siemens controls to ensure reliable, efficient and precise handling of workpieces. Our new Robot Control Multiprocessor family will expand robot capabilities even further into the most sophisticated application areas'. Advertisement in *Fortune*, 18 August 1986, p. 22.

technologies. The advances in modern biotechnology are much more recent than those in microelectronics. As a consequence, few products manufactured with the help of genetic engineering have reached the market yet. Most developments are still in the research field. Therefore, other indicators will have to be used than in the case of applications of microelectronics. While biotechnology is a very different field, there is still some overlap with microelectronics: 'bioinformatics', which encompasses a broad field from expert systems related to databases to special computational devices for prediction of macromolecular properties and special equipment for the computer aided design of proteins. While the position of European research on graphics and computational software on biomolecular problems is regarded as strong and to compare favourably with both the USA and Japan, European manufacturers play hardly any role in the instrumentation commonly used in this field.[24]

Although Europe has a long tradition of 'classical biotechnology' and therefore has a support industry able to supply the equipment for the large-scale application of biotechnology processes, it lacks producers of the most advanced electronic equipment for gene engineering. A good example of the American lead in the instrumentation field is the advance in synthesizer technology. Automated DNA synthesis speeds up the research process. Whereas there are several small American firms offering synthesizers (one of which, Applied Biosystems, founded in 1981, is said to have about 75 per cent of the world market),[25] there are only three manufacturers in Europe. The supply sector in several European countries consists mainly of subsidiaries of *American* companies.[26]

The lack of a strong supply sector is one of the weaknesses of the European biotechnology industry that has brought researchers from the Office of Technology Assessment of the United States Congress to the conclusion that 'the European countries are not expected to be as strong general competitors in biotechnology as the United States and Japan'. According to their study, 'the Federal Republic of Germany, the United Kingdom, Switzerland, and France lag behind the United States and Japan in the commercialization of biotechnology. The European countries generally do not promote risk-taking, either industrially or in their government policies. Additionally, they have many fewer companies commercializing biotechnology'.[27]

Later on in the study, however, the authors acknowledge that 'although there seem to be fewer European companies than Japanese companies

[24] Berendsen (1985), p. 41.
[25] *Le Monde*, 11 June 1984.
[26] *Industriemagazin* (Düsseldorf), March 1983, p. 42.
[27] OTA (1984a), p. 21.

commercializing biotechnology, the potential of European pharmaceutical companies such as HOECHST (FRG), RHÔNE–POULENC and ELF–AQUITAINE (France), ICI, Wellcome, and Glaxo (UK) and HOFFMANN–LA ROCHE (Switzerland) is impressive'.[28]

European experts found that the OTA study does not do justice to the efforts of European companies, and that the information that it contains on them is sometimes blatantly wrong.[29] It is true that European multinationals did not invest in modern biotechnology until after 1981. But they probably were not much later than American or Japanese multinationals. The spectacular progress of American genetic engineering was not the result of research by large companies, but the achievement of small venture companies, closely related to the universities, that popped up in the second half of the 1970s and the first years of the 1980s.

Compared to that in microelectronics, the record of European multinationals in biotechnology is not so bad. The first commercial product of gene splicing was brought to the market in 1982 by an AKZO subsidiary (two vaccines against diarrhoea of cattle and pigs). NOVO in Denmark and GIST-BROCADES in the Netherlands hold dominant positions in enzymes, while the world's third largest producer of enzymes is an American subsidiary (MILES) of the German BAYER concern. ICI has an outstanding position in fermentation technology. UNILEVER has a prominent position in the application of tissue culture. HOFFMANN–LA ROCHE and CIBA–GEIGY are strong in pharmaceutical and agrochemical biotechnology research. But it is true that the largest companies in the European chemical industry play a less prominent role.

To some degree, they have probably been the victims of their own success. The very fact that they have a strong position in the world chemical and pharmaceutical industry can be responsible for their hesitant engagement in a development which is based on a different kind of knowledge and which could undermine their position in the long run.

The Commission of the European Community claims that Europe has the scientific knowledge, the industrial capacity and the agrarian base to be 'the world's number one' in biotechnology.[30]

Europe has *a strong position in basic research*. It is generally acknowledged that basic research in Europe is not less advanced than it is elsewhere. It is even said that with regard to Japan, Europe (together with the United States) has an advantage of two to three years in the fundamental science of gene splicing.[31] European companies have

---

[28] OTA (1984a), p. 110.
[29] See Sharp (1985b).
[30] *The Economist*, 3 September 1983.
[31] *Journal of Commerce* (New York), 11 August 1983.

considerable experience (and a *strong world market position*) *in 'classical' biotechnology*, i.e. in the brewery and dairy industries. They therefore use important *process knowhow* for the large-scale application of biotechnology processes. Many of the world's largest chemical and food producers are European companies. The German and Swiss chemical and pharmaceutical companies possess a very strong position in the world market, and the same can be said with regard to food companies such as UNILEVER and NESTLÉ. Even if these companies have not been in the front line of the development of biotechnology in the past, it can be expected that their very *strong market position will help them to achieve a prominent position* in the commercialization of biotechnology products in the future.[32]

The *Common Agricultural Policy* (CAP) of the European Community, with all the problems connected with it, has made the EC a surplus producer of agricultural products. These surpluses can be used as industrial inputs for biotechnology processes, if the price level becomes attractive enough for the biotechnology industry. Government spending on biotechnology is increasing. Government support for the development of biotechnology does not equal the amounts spent in the United States, but it is higher than in Japan. European (EC) governments together spend about three times as much on biotechnology as Japanese authorities (see Chapter Six).[33]

However, there is another side to this European coin as well. Europe has a strong position in basic research but it has traditionally been very difficult to turn the results of basic research into commercial products. The social distance between academia and business is still very strong in France and to a lesser degree in the UK. It has always been less in the case of Germany, Switzerland and Sweden.

With regard to applied research, it is often difficult to differentiate between American and European research, because many European companies do much of their biotechnology research in the United States. The very fact that they find this a more stimulating environment for their research is in itself a negative indicator of present and future competitiveness in biotechnology.

A strong position in 'classical' biotechnology and process knowhow is clearly an asset. But it caused several European companies to invest considerable amounts of money in fermentation projects in the early 1970s, while they neglected genetic engineering (which turned out to be decisive in order to make the fermentation processes profitable). As a result, several important companies burned their fingers with biotechnology rather

[32] OTA (1984a), p. 110.
[33] *The Economist*, 3 September 1983.

early, with the consequence that 'biotechnology' became anathema at board meetings for a number of years.

Even a strong market position can be a mixed blessing. Companies may show little inclination to develop products that would compete with, or could even be substituted for, their own products on which their market success has been based. Instead, their strong market position may stimulate them to opt for a wait-and-see attitude with some neglect of their own R&D efforts, confident that they will be able to buy up whatever turns out to be viable and interesting.

While the Common Agricultural Policy has led to agricultural surpluses that could be used as feedstock for fermentation processes, it has at the same time led to a price level that is disadvantageous to European producers. European multinationals such as ICI, therefore, have moved large-scale fermentation processes to countries outside the European Community in order to profit from the lower price level of inputs.[34]

The strong power of the European sugar lobby has prevented a large-scale production of isoglucose, which has become the most important product manufactured with the help of immobilized enzymes in the United States and Japan. Although European companies produce most of the enzymes used in this process, they profit less from the equipment demand which results from the large-scale production of starch-based sweeteners.

Another factor that can be very important for the international competitiveness of European biotechnology is the regulatory framework.[35] Since modern biotechnology can generate new forms of life, the debate on regulation is much more intensive in the field of biotechnology than those of other technologies (with the exception of nuclear technology). Where the American Government has taken the position that existing legislation is adequate, the Biotechnology Regulation Interservice Committee (BRIC) of the European Commission is still assessing the adequacy of the existing regimes. The result will probably be that the existing European regulatory framework is sufficient for some pharmaceutical applications of biotechnology, but that new legislation will be necessary in fields like the deliberate release of genetically engineered microorganisms and waste treatment.

The uncertainty about future regulation might be a disadvantage for European industry, but 'representatives of US industry' fear as well 'that other industrialised countries will set up regulatory schemes before agreement is reached among the US authorities, and that this could lead

---

[34] See Lewis and Kristiansen (1985), p. 571.
[35] Roobeek (1986a).

to competitive advantage for foreign companies, particularly if regulation in those countries is less severe than in the US'.[36]

Existing regulations for the introduction of new pharmaceutical and food products do not put Europe at a disadvantage. On the contrary, American companies become active in Europe in order to profit from the sometimes more liberal regulatory framework (or its less bureaucratic implementation) than that in the United States. In addition, the 'step-by-step' approach envisaged by European authorities to master the unknown risks connected with the further developments of biotechnology may be better able to provide detailed information on a broad range of biotechnological products and processes than the 'case-by-case' licensing procedure followed by US regulatory agencies. In the long run, this may imply a higher public acceptance of biotechnology than in the United States.[37]

Since biotechnology is a highly differentiated field (much more than, for example, the telecommunications industry), it is even more difficult to arrive at a comprehensive conclusion with regard to the competitiveness of European multinationals. The fact that no general trends are discernible (in contrast to the semiconductor industry, for example, where a general backwardness had to be acknowledged) in itself is remarkable. The competitive position of European companies in biotechnology related fields seems to be considerably better than in the field of microelectronics.

5.3. THE SOCIAL SETTING

Not all determinants of international competitiveness are company or sector specific. Many depend on the social fabric as a whole, on a country's social structure, traditions, culture. Some of these aspects that are especially relevant to high technology industries will be dealt with in this section. In order to realize innovation processes, a specific combination of scientific knowledge, capital, labour and management capacities are necessary. Each of these four factors will be dealt with in turn. In order to boost innovation, (a) scientific knowledge and technological knowhow have to be transferred from the universities to industry; (b) capital has either to be accumulated by the companies or to be provided by outside sources; (c) labour has to be capable (educated), motivated, available at the right place (mobility) and at an acceptable price (wages). This is

---

[36] Roobeek (1986a), p. 75.
[37] See Roobeek (1986a), pp. 76–8.

not only true for rank and file workers, but all the more for all echelons of management.

5.3.1. The university–industry relationship

With the increasing importance of technology for international competitiveness, many efforts are made to make the borderline more permeable between higher education and public research institutions on the one hand and industry on the other. The reduced life-cycle of products necessitates that basic and applied research are linked more effectively and quickly lead to the development of new products. The efforts of European multinationals to make better use of university research teams to explore risky avenues of development have already been mentioned in Chapter Three. The speed with which scientific knowledge in the institutions of higher learning can be made available for industry depends on a number of factors such as:

—the interest of faculty members not only in basic but also in applied research;
—the acceptability of research contracts from industry (with the eventual implication of secrecy and the timing of publications);
—the opportunity for faculty members to engage in consultancy for private enterprises alongside their teaching and research obligations;
—the relative status of academic and entrepreneur which can be decisive for faculty members' decisions to found (or join) a start-up company to commercialize their findings;
—the degree of risk aversion and entrepreneurial spirit among academics;
—the intensity of interaction between academics and industry researchers in professional associations, advisory committees and the like;
—the extent of grant and visiting scholars programmes of major companies;
—the intensity of regional cooperation of industry and research institutions in 'technology parks', and the like.

The United States is 'distinguished by the range and importance of industry–university ties'. In Japan, 'informal links—based on the 'master-pupil' relation which binds former students to their professors'—play a more important role in university–industry cooperation.[38]

---

[38] In Europe the situation is 'extremely varied, with the closeness of university-industry links distinguishing the "high-performance" countries (i.e. West Germany, Sweden, Switzerland and the Netherlands) from the rest [especially France and the

Recently, most European governments have taken measures to stimulate all forms of cooperation between universities and industry in order to mobilize the inherent innovation potential. Some officials have stressed the importance of applied research with a commercial value to such an extent that even industry fears that the universities will no longer fulfil their intrinsic function in basic research any longer. Cuts in the public budget for universities have convinced many researchers that they should look for additional funds from industry and market their research results.[39] The climate is changing accordingly in Europe. In the case of biotechnology, the links between industry and universities are probably not less close than anywhere else, although they take other forms than in the United States. Fewer university researchers are inclined to open new ventures (given the different industrial structure and the differences in status of academics in the United States and Europe), but that may be as much an advantage as a disadvantage for university–industry cooperation. A researcher contemplating creating a new company probably will be less open to the possibility of sharing the research findings with private companies than one who would never consider such a step.

5.3.2. The cost of capital

Competitiveness is not a question of research results alone.

> The rate at which research results can be put to use depends on how quickly a firm can modernize its plant and equipment to incorporate new products or processes. Thus the effectiveness of R&D in raising productivity is greatly enhanced if the cost of capital is low enough so that business can afford to undertake more R&D and to embody R&D results more quickly in the capital stock.[40]

As far as cost of capital is concerned, Western Europe lies somewhere in between the United States and Japan. The United States has a comparatively low saving rate (6.5 per cent of GDP in 1972–1982), against 20.1 per cent in Japan during the same period.[41] The cost of

---

United Kingdom]. Links between industry and the research system in Sweden have, in fact, been so strong that concern has periodically been expressed about their impact on the independence of the tertiary institutions'. Ergas (1984a), pp. 22–4.

[39] See 'Pushing Europe's scientists out of the ivory tower', *Business Week*, 9 April 1984, pp. 44–7.

[40] President's Commission on Industrial Competitiveness (1985), Vol. II, pp. 107–8.

[41] Op. cit., p. 116.

capital services for the United States and Japanese non-financial corporate sectors was 9 per cent and 16 per cent respectively.[42] Technological development in Japan can profit from the fact that banks are prepared to accept even lower interest rates for investments in technologies that belong to the technologies targeted by MITI.[43] Japanese companies, therefore, work with a very low share of their own capital (5–10 per cent), whereas American companies are highly capitalized (about 70 per cent), with Europe again being somewhere in between (with own capital financing about 50 per cent of total assets). This has important implications for company strategy, — with Japanese companies being more growth oriented and American companies being more profit orientated, while European companies seem to compromise between both objectives.

Cost of capital cannot have been an obstacle for the large European corporations. Many of them had very high cash balances during the early 1980s.[44] But they were probably less challenged by small start-up companies. For these new firms, it has certainly been more difficult to raise capital in Europe than in the United States. During the four years 1981–4, the number of venture investments in Europe was just under half that in the USA, and the average size of individual investment was less than one third. It was estimated that ECU 5–7 billion of venture capital was available in Western Europe at the end of 1984, against about ECU19 billion in the United States.[45]

It is questionable, however, whether this has created obstacles to the creation of new high tech ventures. It seems to have worked the other way round. A 'general lack of sufficient good quality projects or young companies requesting investment' dictated 'that a large proportion of the funds' of venture capital groups were 'channelled into mature companies requiring development capital'.[46] There is a certain interaction process. It has been described above that most of the small high tech (semiconductor) firms in Europe have either been created by government or with strong backing from multinational companies and therefore have not had financing difficulties. Scientists leaving the universities to set up their own ventures to commercialize their findings have remained rare. There was relatively little demand from individuals for venture capital, and this lack of demand probably is an important reason why relatively few venture capital organizations have come about.

[42] Commission of the European Communities (1985c), p. 4.
[43] Bolwijn and Brinkman (1985), p. 15.
[44] Welzk (1986).
[45] Commission of the European Communities, EUR 10224, 1985.
[46] Commission of the European Communities, EUR 10224, 1985.

Table 5.11   R&D resources per researcher in 1981 in business enterprises

	Expenditure per researcher	1971 = 100	Supporting staff per researcher
USA	$104 000	100	n.a.
Japan	$80 000	122	0.90
EC*	$149,000	109	2.54
Other OECD	$117 000	99	2.00

Source: OECD (1986b), p. 24.
*Excluding the UK and Greece.

5.3.3. Human resources

High tech industries are not all capital intensive but, by definition, they are all 'human capital' intensive. It proves to be more difficult to obtain the total number of researchers employed in different OECD countries than to obtain figures on R&D expenditures. The OECD estimates that the number of European researchers is only three fifths that of the United States. Japan, however, has fewer: only about three quarters of the number of Western Europe, if similar definitions are used. But in Japan as well as in the United States, the number of researchers in relation to the total labour force is much higher (higher than 6 per 1000). Only in one of the European countries (West Germany) does the ratio approach 5 per 1000—in all others, it is considerably lower.[47] In addition, research expenditures per researcher (including support staff and equipment) are rather low in Japan compared to European countries, as Table 5.11 shows.

The American system of scientific and technical innovation excels by its diversity (more than 2500 post-secondary institutions, of which some 200 grant PhDs), its comprehensiveness (a large share of higher education in total educational enrolment) and its high level of expenditure per student. Still, American companies perceive a lack of qualified personnel more often than European companies[48] as a constraint preventing a change in emphasis in R&D. But the cause of this difference may not be a less sufficient number of trained personnel, but an indication that the pace of innovation is faster in the USA so that the need for additional or better qualified personnel is more pressingly felt, or that other obstacles are felt more immediately in Europe.

The pace of innovation does not depend on the quality of scientific and technical personnel only. It also depends on the training of the bulk of

---

[47] OECD (1986b), pp. 22–3.
[48] Patel and Pavitt (1986), p. 55.

the workforce. At apprenticeship level, the US system of vocational eduation 'has given rise to considerable concern and may account for some of the US current difficulties in translating a dominant position in *innovation* into dominance in *manufacturing*'.[49] In this field, a few European countries show a better record: West Germany and Switzerland have outstanding systems of vocational education with a combined annual intake of apprentices nearly as great as that in the remainder of the OECD area.[50]

In general, there is a 'closer fit between training and industry requirements in the "high innovation" countries in Europe than in those with a poorer innovation performance'.[51] The lack of personnel with specific technical expertise is regarded as a major obstacle to the use of microelectronic components in manufacturing in France, West Germany and the UK.[52] This reason is mentioned more often in Germany than in France or Britain, which indicates that the need is even more felt (or better articulated) in regions which know the advantages of a highly trained workforce.

It is not only necessary to have a well-educated workforce at all levels, but it is also necessary to have them there where they are needed. That is, the workforce must be *mobile* enough. Labour mobility is far higher in the United States than in either Japan or the EC, with the figure for Japan still being higher than the EC average.[53] In a given year, more than 25 per cent of the labour force changed jobs in the United States, but only about 11 per cent in Japan and 10 per cent in the European Community. Again, the European average conceals very large differences from country to country. Whereas about 19 per cent of the German workers changed jobs and 13 per cent in Sweden, only about 6 per cent of their British or Italian colleagues did so.[54] Very high labour mobility as in the United States, however, may not always be an advantage. While it can increase the diffusion of innovation, it at the same time causes companies to incur additional costs of adaptation and additional training. A high volatility of the labour force can also be accompanied by a lack of commitment to the fate of any given company.

In different surveys, labour relations (the importance of trade unions) were regarded as a negligible constraint on innovation,[55] again with

---

[49] Ergas (1984a), p. 19.

[50] 'Moreover, the percentage of completions of apprenticeships is substantially higher in these countries, and the quality of apprentice education is generally considered to be superior'. Ergas (1984a), p. 20.

[51] Ergas (1984a), p. 20.

[52] It was mentioned as the major obstacle in a survey of more than 3500 factories in the three countries in 1983 by Patel and Pavitt, (1986), p. 51.

[53] This in spite of the fact that there is no 'certification' mechanism for skilled labour in Japan, which 'is one factor explaining low labour mobility in Japan'. Ergas (1984a), p. 20.

[54] Ergas (1984a), p. 60, exhibit 26.

[55] Patel and Pavitt (1986), p. 56.

considerable differences between European countries: difficulties with shop floor and unions in the use of microelectronics were mentioned more often in France and in Germany than in Britain.[56] In all countries, however, many other constraints were regarded as more important.

5.3.4. The quality of management

One aspect of human resources deserves special attention—the quality of management. Since most studies on international competitiveness are done by or for industrialists or governments, this is a factor that has not been analysed in as detailed a manner as would be desirable—probably for a number of reasons. It is difficult to analyse and to measure in the first place. It is difficult to change, because middle level managers tend to conform to national stereotypes; shop floor workers and top managers are more similar to their equivalents abroad than middle level employees. Policy oriented studies, therefore, will not pay too much attention to it. And finally, it does not seem opportune politically, since most of the readers of studies of this kind may be more inclined to look for the reasons for a lack of competitiveness in the labour camp rather than the management one.[57]

In the chapter on the social impact of the application of new technologies, we have stressed that one group that will be especially negatively affected by increasing automation and the installation of management information systems will be middle management (see p. 113). The introduction of new computer-based systems alters the power relations between different managerial groups, and this invariably induces resistance from the losers. According to M. Lynne Markus, of MIT's Sloan School of Management,[58] this may be due to three aspects: (a) altering access to information, (b) the system's use in modifying behaviour, and (c) the symbolic aspects of information systems.

(a) The new systems provide central management with information that hitherto was only available to managerial staff nearer to the work floor, which derived some power from this position.

[56] Patel and Pavitt (1986), p. 52.

[57] Margaret Sharp notes in the last lines of her book: 'Much emphasis has been put to date on structural rigidities of the labour market. These studies suggest that while these rigidities play some part in Europe's somewhat laggard response to new technologies, the mote may be as much within the eye of the industrialists and Governments who make these assertions, as within the markets and institutions which they so readily criticize' (1985a), p. 295.

[58] M. Lynne Markus (1981), *Implementation Politics: Top Management Support and User Involvement. Systems, Objectives, Solutions*, vol. I, p. 209, cited in Shaiken (1985), p. 232.

(b) Computer-based systems can take away a lot of flexibility, because detailed day-to-day control becomes possible, and middle managers lose the autonomy to take corrective action before their superiors discover a problem.

(c) The new systems may either deprive middle management of status symbols they hitherto had (e.g. secretaries) or provide shop floor workers with access to the same kind of tool (e.g. terminals) with the result that its value as status symbol diminishes.[59]

Considerable resistance, therefore, can be expected from the employees affected (or threatened). Since this resistance normally does not take militant forms, but often is extremely subtle and consists more of passiveness than of spectacular protest measures, it may have escaped broader attention. The resistance of middle level management may be an important factor in explaining the different performances of German companies (with a relatively small number of echelons between top management and shop floor) and French companies (with an elaborate hierarchy of different management levels, see Chapter Four). But could this factor also be of importance in explaining the lack of European competitiveness in comparison with the United States and Japan?

One significant difference between most European multinationals and American or Japanese multinationals is the degree of mobility of higher employees of all kinds. In the United States, managers typically switch from one company to another during their career. If they do so, their standing only increases. This is not the case in Europe. Here, management staff show less inclination to look for a better position outside their company, and companies are less inclined to attract outsiders to fill management positions. A possible candidate with experience from a large number of companies would probably not be regarded as a good choice, because he would have the image of a 'job hopper' rather than that of a serious manager.

What is the impact of these preferences on international competitiveness? The consequences are two-fold. First, there is less diffusion of experience among companies. The literature on technology transfer stresses that the most effective forms of technology transfer are accomplished by the mobility of people who take their knowhow and experience with them. This diffusion is reduced if the mobility of key personnel remains low.

Second, and at least as important, is the consequences for the internal flexibility of companies. With key people remaining in their position for

---

[59] See the publications cited in Shaiken (1985), pp. 231–4, and notes 18 and 19 on p. 293.

a much longer time, they identify more with the specific division they are working in, and divisions acquire more separate interests that they defend against each other. While some competition between the divisions can have positive effects as well, the conflict often is resolved by allocating divisions a specific (and rather constant) share in the company's total resources (e.g. investment, R&D money). In such a situation, it becomes much more difficult to accomplish a massive shift from more traditional activities to new and risky but potentially more rewarding activities. Strong established interests will defend the position of the company's more traditional branches.[60]

This may help to explain why European companies often excel in rather traditional areas where they perfect design and performance of rather traditional devices and products, while showing a great reluctance to come up with radically new products. That is because established interests of different branches assure a more or less proportional allocation of research funds over different divisions, which themselves are already more numerous and heterogeneous in the case of many European multinationals in comparison to their American counterparts (see beginning of Chapter Five).

If this is a valid agreement, how, then, can it be explained that Japanese companies in many cases perform much better, although intercompany mobility is even less than in Europe? The answer is simple. While intercompany mobility of higher employees is much less in Japan, where life-time employment with the same company is the rule (as far as the large companies are concerned), *intracompany mobility* in Japanese companies is much higher. Most companies even follow the rule that no individual keeps the same position for more than one year. Rotation of personnel between production and engineering and product development, for example, is normal practice in 61.3 per cent of large Japanese companies, but only in 27.4 per cent of European or 14.1 per cent of American companies.[61] As a result, people develop less loyalty to specific branches and divisions of a company and more loyalty with regard to the company as a whole (with the side-effect of the creation of dense networks of intracompany contacts which facilitates the diffusion of experience and knowledge, increases mutual understanding and smooths the decision-making process on common priorities). Therefore, Japanese companies can be expected to be better able than European companies to target their

---

[60] Company policy and organization constraints, accordingly, are seen more often as constraints preventing a change in emphasis in R&D in Europe than in the United States. See Patel and Pavitt (1986), p. 55.

[61] Answers to questionnaires filled in by 200 Japanese, 164 European and 177 American large companies; see Meyer (1986), p. 87.

R&D efforts on specific goals, instead of spreading the resources widely over a large number of different activities.[62]

5.4. CONCLUSION

This chapter has provided a mixed picture of the competitiveness of Europe and European multinationals in comparison with companies from the United States and Japan. This comparison corroborates the analysis of Margaret Sharp that the actual technology gap, just as the 'technology gaps' discussed in the 1950s and the 1960s, is more a gap in the *commercialization and use* of new technologies, is more a management gap, than a gap in technology *per se*.[63] It has become obvious, again, that differences between European countries are often larger than differences between Europe as a whole and the United States or Japan. In many cases, the situation in a few European countries (notably West Germany, Sweden and Switzerland) is rather similar to that in the United States or Japan, whereas other European countries show a record which is much worse.

The large differences in competitiveness between different European companies and countries can explain many differences in the government–business relationship across Europe to which we shall turn in the next chapter. Given the rather favourable position of West Germany, it is not astonishing that a relatively liberal policy has been followed there. What cannot be explained on the basis of the analysis, however, is the very market oriented British policy. To explain this, quite another type of analysis is needed.[64] The enormous differences inside Europe also help to explain the difficulties in coming to a viable. cooperation among European companies, which will be dealt with in Chapter Seven.

---

[62] Japanese researchers, nevertheless, have shown considerable capacity to improve the design of even the most traditional instruments and devices — such as a hammer, for example, whose *Durchschlagskraft* could be enhanced without increasing total weight by introducing a mobile mass into the hammer's metal head, gaining some additional kinetic energy.

[63] Sharp (1985a), p. 291.

[64] Henk Overbeek, dissertation (forthcoming), Amsterdam.

Chapter Six

High Tech Multinationals and National Innovation Policies

Increased international competition as a result of the worldwide economic crisis in the 1970s has led to an intensification of government intervention in almost all European countries. For the reasons described in Chapter One, national governments (as well as the European Community as such) have taken a much more active stance with regard to industrial and innovation policy—to such a degree that one can even speak of a 'restructuring race' among the major capitalist industrial countries. Governments have become so intensively involved in the promotion of national industry that international competition between companies to some extent has turned into outright rivalry among states.[1]

In this chapter we analyse how far increasing state intervention has led to a shift in the balance of power between the leading European multinationals and their respective national governments. Have European multinationals become more dependent on public subsidies, and is technological development more and more steered by state bureaucracies rather than management? Or have governments become increasingly dependent on the high tech multinationals in order to formulate and execute their innovation policy? Have national governments and their 'national' companies become more independent, or has this relationship become less close as a result of the internationalization process?

The relation between the national state and multinational firms is the outcome of a continuous bargaining process. In this process, each actor has its own set of objectives and instruments. There is no doubt that governments and multinationals have many interests in common. Governments are interested in an economy with a strong technological capacity which is internationally competitive in order to retain a certain national autonomy and generate an acceptable level of general wealth. The productivity of the national industry in the longer run influences the political margins of any government. The economic and military strength (which depends very much on technological sophistication) decides whether a state will be an important actor in world politics as well. Companies

---

[1] Junne (1984).

share the interest in a prosperous economy in so far as it generates demand for their sophisticated products. They share the interest in a high level of education and in a developed infrastructure, which can reduce their costs of production and commercialization.

But whereas national governments try to boost activity and income within the national boundaries, in the first place, multinational companies care much less where the input (labour, knowhow, intermediate products) and the demand for their products come from. And whereas they share the interests of the state bureaucracy in specific government outputs (infrastructure, subsidies, public demand), their interests with regard to the financing of these outputs (corporate taxes) can be diametrically opposed. It is because of these divergences of outlook that an analysis of the possible shift of balance in the relationship between governments and multinationals as a result of accelerated technological development and speeding up internationalization is of importance.

In dealing with the relation between multinationals and national states we have to discern between multinationals originating from large countries and those from small countries. Governments of smaller states in general have less bargaining power than those of larger states. The same is true of the bargaining position of larger versus smaller firms. If large firms in a small country place their laboratories elsewhere (as has often been threatened) this would have a considerable impact on the small state's power base: it might instantly be stripped of 10–20 per cent of 'its' research capacity (see Table 6.5). While smaller countries more often offer a liberal environment and an advantageous tax system, larger industrial states in general have more active support to offer (outright subsidies, public procurement), and are less affected by individual multinational firms threatening to transfer their activities abroad (it might ultimately cost them 1 or 2 per cent of the national R&D capacity).

The intensity of state support will vary over time, and it will vary with the position of the state in question in the international division of labour. Variations over time can largely be explained by changes in the overall economic climate. In times of economic prosperity, governments will be more inclined towards a *laissez-faire* attitude than in times of economic crisis.

The relationship between government support and a country's position in the international division of labour is less straightforward. Countries with a top position in the international division of labour do not have to support 'their' companies intensively. During the three postwar decades, the United States, for instance, did not use her domestic powers to the full extent or in a more mercantilist way because of the hegemonic position of her national industry. Countries that want to 'catch up' (like Japan,

France) on the other hand show more direct government involvement. Countries in a disadvantageous position without much of a chance to take an active part in technological developments again will have less active government support (and may try to participate in developments originating elsewhere by following a *laissez-faire* policy instead).

This pattern, however, seems to have changed as a result of the economic crisis and increased international competition. There are no clear-cut top positions any longer, and the American industry has come under increasing competitive pressure as well. As a result, even the Federal Government of the United States is extending its reach more and more so that the USA has been considered even more protectionist than Japan.[2]

In this chapter, first the international subsidy race will be described (section 6.1). Next, it is shown to what extent government support concentrates on the large companies (section 6.2). In spite of all the internationalization, the relationship between national governments and multinationals, especially in the field of electronics, turns out to be still very close. In the third part of the chapter it is analysed how governments promote national industry by changes in the regulatory framework (section 6.3).

6.1. THE INTERNATIONAL SUBSIDY RACE

The history of multinationals, in the information technology cluster more than anywhere else, is closely intertwined with national state intervention. State intervention in general increased dramatically after the Great Depression of the 1930s in all capitalist countries, be it in fascist Germany or Italy or under the Popular Front in France. The closest interaction between governments and companies was reached during the Second World War, when the activity of all companies was geared towards military needs. Systematic stimulation of technological innovation by governments took shape during the war, and many government–business links continued afterwards.

[2] The USA has approximately 45 per cent of its manufactured imports subject to some form or another of non-tariff barriers whereas Japan has only 22 per cent of its imports subject to restrictions. *Business Week*, 8 April, 1985. It has to be noted that there are other less formal barriers which might make the outlook for Japan more protectionist. What is important, however, in this context is that the validity of the traditional distinction between the US as liberal and Japan as protectionist is open to discussion.

6.1.1. Government intervention before the 1970s

After the Second World War, the pace was set by the US Government, which used its defence budget to encourage important breakthroughs in semiconductor technology, computers, telecommunications equipment and advanced machine tools. The influence of the Pentagon was strongest during the 1950s and 1960s when the Cold War and the Sputnik shock offered a favourable ideological climate. The success of many large US electronics firms has been closely linked to their relationship with the Pentagon. IBM, for instance, was only one of many firms producing calculating and registering machinery[3] in the first postwar decade. The winning of important Pentagon contracts made it possible for IBM to develop new generations of computers on the basis of secure procurement contracts and R&D cooperation with the military. These military contracts provided the firm with a base from which it could more easily market its large computers to the civilian sector.[4] This support strategy strengthened the firms involved, and helped them to expand internationally to penetrate the economies of other industrial countries. The massive support given to those American firms made it almost impossible for other countries to retain a presence in (micro)electronics without state intervention.

It was not until the dominance of the American multinationals became overwhelming that European governments became more actively involved in the support of a domestic technology base. The reaction of European countries came later than the Japanese.[5] The relatively late reaction in Europe is probably caused by the dominance of American multinationals in Europe, more open economies and the political priority attached to the rebuilding of the steel, automobile and shipbuilding industries. Only after the mid 1960s did European Governments react by supporting parts of the (remaining) electronics industry. In 1967 West Germany began to support its computer and microelectronics industry.[6] The 1967 'Plan Calcul' in France also subsidized the electronic components industry. The French strategy was inspired by the refusal of the US Government to provide an export licence for mainframe computers to France, which were needed for the French nuclear programme, thus frustrating French nuclear aspirations.

---

[3] See Freeman (1982), pp. 86, 87.
[4] See Sobel (1983), p. 99.
[5] Japan resorted to direct protection and intervention as early as the 1950s. In 1958, for instance, an indigenous Japanese computer industry was started with government-sponsored research; support of the machine tool industry dates back to the mid 1950s; until the 1960s only IBM and Texas Instruments were allowed to settle in Japan.
[6] US Department of Commerce (1983b); Dosi (1981).

Table 6.1 *Major structural changes and government intervention in the electronics industry in the UK, France and Germany, 1968–84*

FRANCE

1968	1970	1973	1975	1978	1978–9	1981–2	1983–4
Government fosters SECO (Thomson) and COSEM (CSF) merger to create SECOSEM (Thomson) heavily supported by state	EFCIS (semi-conductors) created as joint venture between Thomson and CEA (Atomic Energy Commission)	Creation of UNIDATA (computers), a joint venture of CII, Siemens and Philips. Uncertain government attitude	Failure of UNIDATA. Government supports merger of CII and Honeywell-Bull	Thomson takes over semiconductor division of LTT and SILEC	Government supports joint ventures of: Saint-Gobain/National Semi-conductor; Matra/Harris; Thomson/Motorola. Saint-Gobain entry into CII and Olivetti. Support of Radio-Technique (Philips). (5 poles of production)	Nationalization of CGE, Thomson, Saint-Gobain, CII-HB (becomes Bull); majority stake in Matra	Concentration of computer activities of Thomson, Saint-Gobain and CGE with Bull. Thomson takes over: the joint venture of Saint-Gobain and National Semiconductor (Euro-technique) and the semi-conductor business of CGE. Of the 5 poles of production, only two remain. Saint-Gobain withdraws from Olivetti, CGE takes a 10% share in the Italian firm

GERMANY

1970	1973	1975	1978–9	1979–80	1983	1984
Creation of DATEL. Joint venture of state, Siemens, AEG-Telefunken, Nixdorf (in computer applications)	Creation of UNIDATA (see above). Favourable government attitude	Siemens takes over big computer division of AEG (approved by state)	Rescue of AEG-Telefunken by a consortium of banks. Indirect Federal support	Plans for the establishment of a joint research laboratory of three major firms and public agencies (Berlin Synchrotron Projekt)	Semiconductor division of AEG merged with Mostek (United Technologies) in a joint venture. Telefunken taken over by Thomson	Joint research in Germany of ICL, Siemens and Bull (in computers and information technology). Takeover of Grundig by Philips (after disapproval by the Bundeskartelamt of the same effort by Thomson)

UK				
1968	1976–8	1978	1980	1984
Joint venture between Mullard (Philips) and GEC, taken over by the former. Series of mergers lead to ICL (computers) 10.5% owned by state	NEB buys shares in Ferranti (computers, semiconductors, military, etc.) and in various small and medium firms in software, industrial and consumer electronics	Constitution by NEB of INMOS (VLSI memories and MPUs). Entirely publicly financed	Conservative Government sells ICL and Ferranti to private market	Government sells its 75% share in INMOS to Thorn–EMI. STC (25% owned by ITT) tries to acquire ICL

Source: data 1968–80 from Dosi (1981), p. 94; after 1980 from own observations.

Table 6.2   Subsidies to private enterprises as a percentage of GNP

	1970	1982		1970	1982
Belgium	1.3	1.6	Denmark	2.8	3.2
France	2.0	2.2	West Germany	1.7	1.8
Great Britain	1.7	2.0	Italy	1.5	2.9
Netherlands	1.3	2.5	Norway	5.1	6.5
Austria	1.7	2.9	Sweden	1.6	5.0
Switzerland	0.8	1.3			

Source: Bundesministerium der Finanzen (1985), p. 20.

In the UK, the government intervened directly in order to merge dispersed computer activities into one firm, thereby creating ICL in 1968. Table 6.1 illustrates for the UK, France and Germany how the structure of the electronics industry has been intimately influenced by direct government intervention.

6.1.2. Stimulation of innovation as a response to the economic crisis

Reduced economic growth, high inflation rates, increasing unemployment and large balance of payments deficits caused most industrial countries in the second half of the 1970s to intensify their industrial policy. The OECD speaks in this context of a 'new interventionism' as a result of the economic crisis.[7] As a result, subsidies to industry increased considerably (see Table 6.2).

While there has been an overall increase in subsidies to companies (especially in the smaller countries), a shift has taken place in the distribution of subsidies, with subsidies to 'new', high tech industries becoming more important. During the first years after the 'oil shock' in 1973, governments had given strong support to declining industries. But when it became obvious that the crisis was structural and that some traditional activities could not be rescued, governments increasingly looked for new sectors to promote. In most cases, this support did not replace the support to ailing sectors, but was added to it. Most of the increase in subsidies went to the 'new' sectors, which could thereby increase their share in total subsidies.

For the reasons described in Chapter One, all countries concentrated their support on more or less the same sectors and technologies, although

[7] OECD (1983c), p. 48.

the instruments chosen varied from country to country. Three groups of countries show notable differences,[8] which can be related to whether they form the home base of important multinationals or not.

First, we have the large European countries — France, Germany, the UK (and Italy, to some extent) — which try to maintain a position in *all* core technologies. Their governments emphasize very much the importance of national programmes and hesitate to agree (or even obstruct) a substantial expansion of the European research and development programmes.

Next come three smaller countries — Sweden, Switzerland and the Netherlands — which are comparatively large R&D spenders (measured by R&D expenditures as a percentage of GNP) and have a share of R&D personnel comparable to the larger countries. These are the countries with a large number of home-based multinationals. The R&D performance of the third group, the other small countries with no major home-based multinationals, is generally lower.

In many instances, the policies of the three 'larger–smaller' countries run parallel with the group of the larger countries due to a comparable interdependence of domestic multinationals and government. In the larger countries and this group of three countries, government-financed R&D (as a percentage of total government expenditure) has, after a long downward trend, followed by a levelling off, 'generally been surviving budgetary restraint better than the bulk of other government activities.[9] In other medium and small OECD countries this ratio was lower and was stagnating or declining in the period up to 1983.

6.1.3. Institutional changes

The increase in the amount of subsidies went along with institutional changes in government bureaucracy. Whereas most subsidy programmes in microelectronics originally were initiated by the Ministry of Economic Affairs (or Industry), Postal Services or Defence, other ministries (e.g. Ministry of Education) increasingly became subsidizing agencies as well since it became clear that the race in core technologies was a race in diffusion and application as well as in production.

---

[8] See also the distinction made by the OECD (1986b) in its publication on science and technology indicators between 'major', 'medium' and 'small' R&D countries.
[9] OECD (1986b), p. 29. The OECD only considered R&D data until 1981–3. In the years after not only the Dutch and Swiss expenditures, as noted by the OECD, but also the Swedish expenditures have risen considerably.

This raised the question how an integrated industrial and science and technology policy could be managed. In many countries this has been a subject of conflict between departments not only regarding who is to coordinate and to finance the large technology projects.[10] The outcome of these conflicts also partly decides who is going to benefit from the support schemes, given the different constituencies and traditions of different ministries. In most countries this discussion has been linked to the question whether to found a separate Technology Ministry. The smaller countries provide an example of fairly stable institutional settlements with only gradual changes. The grip of the Ministry of Economic Affairs in the Netherlands on innovation policy, for instance, gradually became more intensive with the result that it was able to raise its share in all government expenditures on R&D from an average 10 per cent in the first half of the 1970s to 20.1 per cent in 1985.[11] This shift has been explicitly supported by general government advisory committees in which the large home-based multinationals are strongly represented.[12]

Germany is the only larger European country with a relatively stable institutional setting for innovation policy in general and support of microelectronics specifically. This is partly due to the early creation of a Federal Ministry for Research and Technology (BMFT). In 1972 this ministry was created to promote industrial change through the use of public R&D funds. As a result, German technology intensive industry had a focal point in the BMFT for funding and support within the government bureaucracy from the early 1970s on. In the other large countries most high tech industries had to strive for funding from traditional ministries which more often than not were still dominated by the direct interests of mature industries like shipbuilding, steel and textiles.

As a result of the growing internationalization of industry and growing concern for international competitiveness, governments in the larger European countries tried to adapt the institutional setting to integrate industrial policy with trade policy following the highly successful Japanese MITI approach. In Italy, France and the United Kingdom Departments of Trade and Industry were created in the early 1980s. They provided important support from within the government bureaucracy for lobbies against policies affecting the competitive international position of the national industry (especially with regard to a too-stringent implementation of antitrust laws; see also section 6.3).

Both Britain and France experimented with restructuring their political institutions in order to create separate political responsibility for technology.

---

[10] Rothwell (1985), p. 1.
[11] OECD (1986c), p. 15.
[12] Rathenau Commission on electronics; Wagner Commission on reindustrialization, Pannenborg Commission on public procurement and innovation.

Both countries thereby developed stop–go practices which represent the rather ambiguous technological position of their industry and thus the almost, by definition, limited effectiveness of every policy chosen. France, for instance changed the institutional setting of the industrial and technological policy five times in the 1981–6 period.[13] The competence battle in France over the financial responsibility for the national electronics industry programme — a FF10 billion obligation each year — has been extremely fierce. The socialist administration in 1984 moved it from the Ministry of Industry towards the PTT (without the latter's approval). With the centre-right government, the responsibility in 1986 was transferred to the newly formed Ministry for Industry (and Tourism and PTT).[14]

Whereas in the 1970s the trade unions were involved in the formulation and even in the management of parts of the industrial and technological policy through tripartite organizations, their influence has gradually declined to such an extent that at present policy is formulated — even in many smaller countries — in committees where only government, employers and so called 'experts' participate. By and large 'experts' have taken over the participatory role of trade unions. Perhaps the most prominent example of declining labour union influence in Europe can be found in Germany. The 'Humanization of Working Life Programme' starting in 1974 was a comparatively early effort to try to combine innovation in process automation and quality of working conditions. In the 1974–84 period more than 1000 projects were supported with around DM100 million per year. Although a relatively small proportion of the Technology Ministry's budget (around 1.5 per cent), the programme nevertheless enabled trade union representatives to participate and influence the outcome of the project at all levels. The programme lost its attractiveness for the technology ministry and participating (large) companies as soon as the effectiveness of rationalization in the international technology race became more urgent than to win the trade unions' favour.[15] In fact, the parts of the Humanization Programme which more specifically tackled robotics and flexible automation were transferred to a separate programme for 'Industrial Technologies', with far less trade union involvement.[16]

Another aspect of institutional change was a switch from individual projects to more comprehensive support programmes, into which different projects were integrated. This can best be seen in the field of

---

[13] These ministries were: Ministère de la Recherche et de la Technologie (Chevènement), Ministère de la Recherche et de l'Industrie (Chevènement), Ministère de l'Industrie et de la Recherche (Fabius), Ministère de la Recherche et de la Technologie (Curien), and in the Chirac Cabinet: Ministre délégué à la recherche et à l'enseignement supérieur (Devaquet) together with the Ministère de l'Industrie, des PTT et du Tourisme (Madelin).

[14] FinTech, Telecom Markets, 54/7, 1986.

[15] Dankbaar (1986), p. 40.

[16] Dankbaar (1986), p. 41.

microelectronics, where individual projects to support the development of semiconductors have become included in programmes for information technology, promoting not only research on components, but also their application as well as the diffusion of new products and processes.

6.1.4. Government programmes in microelectronics

Microelectronics provides a prime example of steadily growing government involvement through subsidies. With the exception of support for telecommunications and defence equipment, subsidies in the 1960s and early 1970s were measured in tens of millions of national currencies. The decade after the mid 1970s showed an explosive growth of subsidy programmes for microelectronics for which presently only measurement in billions of national currencies is appropriate. The 'Fillière Electronique' programme in France, for instance, was planned to provide French industry with FF70 billion support in the 1982–6 period. In Germany, the 'Informationstechnik' programme should provide around DM3 billion for industry in the years 1984–8. In the Netherlands an Informatics Stimulation Plan for the 1984–8 period was budgeted at DFL1.7 billion. Despite its anti-interventionist attitude the British Conservative Government nevertheless boosted its spending on projects like the Alvey project on fifth generation computers and advanced components (expenditures for 1983–8: £350 million) to levels considered higher than the German state support for information technology.[17]

Not only the amounts of support have risen, but also the number of support programmes and supporting public agencies have often multiplied. Influenced by the actions of other countries, most countries subsidize the same core technologies. Consequently, the amount of duplication is considerable.

Whether or not support will be given and in what form depends to a large extent on the existence of large firms operating in the area. The fact that Germany and the Netherlands are the home of the two remaining relatively successful European chip producers (SIEMENS and PHILIPS) made support of DM300 million and DFL200 million for the core technology of large memory chips with submicron technology (Megabit) almost mandatory once these two producers threatened to move their research and production facilities

[17] Webber, Moon and Richardson (1984), p. 23.

abroad.[18] THOMSON, against its will, was left out of the German–Dutch project, but successfully lobbied the French Cabinet for approximately FF1 billion in support of its own megachip project.[19] Selective subsidy policies in the core technologies of the microelectronics cluster thus are largely created in response to specific requests by large producers which increasingly become unable, or unwilling as in the case of SIEMENS which has huge capital reserves, to finance the growing investments on their own.

The genesis of the Alvey programme in the United Kingdom, as an explicit reaction to the Japanese fifth generation computer project, provides an interesting sample of how a project's shape can be influenced by firms and different public agencies. The initiative for the project apparently came from BRITISH TELECOM[20] while it was still a state monopoly in order to retain considerable state support for information technology after its privatization. Soon, however, the Ministry of Defence (MoD) also became involved. In the international race in microelectronics the ministry was eager to have a British project in VLSI and advanced computers, fearing that the country would lag behind the USA. The MoD changed the original civilian orientation of the project into more military applications with most research aimed at components research.[21] As a consequence the major military contractors—PLESSEY, GEC, FERRANTI and STC—became also the dominant firms in the project.[22] Alvey was the only programme not affected by a government moratorium on new grants to industry under the 'Support for Innovation' programme in 1984–5. Since most of the large firms are involved in the Alvey programme, whereas many smaller firms are making use of the other innovation grants, this can be regarded as an indicator of the influence of large firms.

The only small country with a mainframe computer industry of its own has been Sweden, after the UNIDATA project had been terminated, in which PHILIPS (supported by the Dutch Government) had cooperated with SIEMENS and BULL. Restructuring became necessary, however, due to the threat of US computer firms. The Swedish Government chose an offensive policy in the second half of the 1970s. It influenced the

---

[18] It seems that Siemens threatened to move its facilities to Silicon Valley in California, whereas Philips influenced the Dutch Government by stating that it would locate all its chip production in Hamburg, Germany. It is interesting to see that if both countries had refused to subsidize, on balance the German Government probably would not have been hurt because the German investment in Philips could have offset the disinvestment in Siemens. The smaller country, however, would have lost in any other but the chosen scenario.

[19] By doing this the company overruled DIELI, the department at the Ministry of Industry which is responsible for dealing with the firm, which refused to support the project arguing that its economic feasibility was disputable.

[20] Mr Alvey is Technical Director at British Telecom.

[21] Interviews in London, 18 July, 1985.

[22] *Social Europe*, December 1985, p. 120.

industry to withdraw from the production of mainframe computers and supported the merger of the two largest Swedish computer firms (DATASAAB division of SAAB–SCANIA and STANSAAB) in order to create cooperation in R&D and in marketing peripheral equipment and special purpose computers.[23] The government took a majority share in the newly created firm. The new company, DATASAAB AB, however, proved very weak and losses of hundreds of millions of kroner had to be covered by the government. This put a heavy claim on the total Swedish budget to support electronics. More than 90 per cent of the total SEK425 million in support of electronics in the 1970–9 period went to DATASAAB.[24] In 1980/1 ERICSSON took the company over,[25] which ended an expensive state experiment.

One of the rare other examples of a small country supporting the development of one of the core technologies in microelectronics without having a major home-based multinational in that area comes from Switzerland where the government tried to develop a national digital public switching system. Under government tutelage cooperation was established between SIEMENS–Switzerland, STR (ITT subsidiary) and HASLER, a Swiss supplier of telecommunications equipment. In 1983, a few years later, after having spent more than SFR200 million, the Swiss PTT ended its effort and decided to buy an already developed switching system (from ITT). An own system proved to be impossible before 1986, especially due to huge problems with the development of advanced computer software. By that time, newer and more advanced systems could have reached the market, leaving Switzerland with an outdated system.[26]

Like Switzerland, many other small countries have seen ITT as their 'national champion' in telecommunications. Austria and Australia stipulated that the switching equipment they bought from SIEMENS and ERICSSON respectively had to be produced under licence by 'their' ITT subsidiary. The function of the smaller country in this respect has been an important consideration in the marketing strategy of ITT towards third countries. In the case of China, for instance, Bell Telephone (ITT Belgium) was chosen to bargain for the contract. The Belgian foreign minister, Tindemans, lobbied successfully not only the Chinese, but also the US Government, to give the order to Bell Telephone.[27] This would never have been done by the

[23] Carlsson (1980), p. 10.

[24] Hingel (1982), p. 20.

[25] Geoffrey Foster, Ericsson's electronic ambitions, *Management Today*, June 1983, pp. 54–5.

[26] *The Times*, 11 August, 1984.

[27] It is interesting that the American Government approved the Belgian mediation only after it became clear that no other American producers were among the suppliers and the competition was left between Bell and CIT-Alcatel (CGE) and Ericcson, i.e. two other European producers!

government of Germany or the Netherlands where ITT is only a second
or third supplier next to the national champions, SIEMENS or PHILIPS.
With the takeover of the telecommunications operations of ITT by CGE,
it becomes a matter of cautious bargaining whether the former ITT
subsidiaries will still be considered as national champions, or more bluntly
as affiliates of a French corporation.

Support for the telecommunications industry does not so often take the
form of outright subsidies for research efforts. The PTTs often do research
on their own, but an extensive exchange of research results is taking place
with the largest national contractors (see 6.2.3). PTTs plan and finance
pilot installations (see Table 5.6), which provide the largest national
telecommunications companies with orders and experience. The sophistication of the equipment demanded depends very much on the quality of
the network. All European governments accordingly have plans to
modernize and extend the national networks not only to promote the
international competitiveness of their national telecommunications
industry, but to increase the competitiveness of the national economy in
general. In order to avoid these national efforts resulting in incompatible
national specifications and standards and thus a further fragmentation
of the European market for telecommunications equipment (and a slower
diffusion of new equipment as a result of its limited reach), the European
Community has started its own telecommunications programme, called
RACE (Research and development in Advanced Communications
technology in Europe). The pilot phase started in 1985. It aims at the
development of common standards for a Europe-wide ISDN network and
connected equipment. All major producers and PTT laboratories
participate in the programme (see also Chapter Seven).

*Integration of core technologies in the automation of
production*

It is not so much the development of individual core technologies which
decides on the degree of international competitiveness, it is rather the
integration of core technologies into systems of production and office
automation. Government support programmes here clearly follow the pace
set by the large firms as the most obvious protagonists and the earliest
users of these systems. In production automation, however, an important
part of the national capability lies with the machine tool industry. This
sector consists mainly of small and medium sized (often family-owned)
firms in most European countries. The success of subsidy programmes,
especially in the larger countries which all started programmes in support

Table 6.3   Direct government programmes to promote automated production

Country	Nature of programme
GERMANY (production and diffusion)	New Production Technology Programme (1984–7: DM530 million); CAD development (DM160 million). Estimated total grants between DM200 and DM300 million per year (Federal plus Länder support).
FRANCE (production and diffusion)	National Machine Tool Plan (1981–4: FF2.3 billion); Fonds Industriele de Modernisation (FIM): aims at mobilizing savings for investment in advanced manufacturing; Automatics and Advanced Robotics (ARA) Programme; etc.
ITALY (production and diffusion)	Mechanical Engineering Technologies Project (follow-up of 1979–84 'Industrial Automation Project): LIT31 billion over 5 years; Special Fund for Technological Innovation: from 1982 appr. LIT3.650 billion; since 1983 Bill N 696 supporting SMEs to buy NC machine tools: 1983–5 period LIT90 billion in funds.
UK (production and diffusion)	Department of Trade and Industry programmes (1981–6): production techniques (£90 million); production and development of microelectronics (£50 million); Microelectronics Applications Programme (£50 million); use of Science and Engineering Research Council in programme on the Application of Computers to Manufacturing Engineering (ACME) £4 million in 1983.
NETHERLANDS (diffusion)	Demonstration projects for introduction of flexible manufacturing systems; 1983 budget around DFL12 million. Stimulation of Innovation (INSTIR) project supports wage costs (DFL1100 million).
BELGIUM (diffusion)	Action plan for microelectronics technology. Robotics budget: BF100 million (total budget: BF2.655 million). CAD/CAM support only for R&D for applications in the design and production of integrated circuits.
DENMARK (diffusion)	Technological Development Programme 1985–9 promotes information technology in main industrial sectors (total budget: DKR1525 billion).
SWEDEN (diffusion)	'Robot-84' campaign; provision of risk capital for demonstration projects and awareness campaign; feasibility studies are undertaken and support for training programmes of around SEK108 million since 1983.

Source: Social Europe, special supplement 1986; own observations.

of both production *and* diffusion of automation technology (see Table 6.3), thus not only depends on the effectiveness of the government–'national champion' relation but also on the wider infrastructure in which the relationship with the (harder to reach) smaller firms is of special importance.

In Sweden, Germany and Italy the machine tool industry succeeded in remaining fairly strong despite the Japanese challenge in (C)NC machine tools. Italy is perhaps more of a European success story than Germany according to Horn, Klodt and Saunders. 'Helped by the major links forged with the United States, the Italians are now in a better position than any of their EEC counterparts to move into full-scale automation.'[28] This conclusion might be an overstatement because of the far better developed and evenly distributed infrastructure in the Federal Republic which enhances the effectiveness of any restructuring strategy that increasingly necessitates a joint effort of business, universities and other governmental agencies. Especially the technical institutes (Fraunhofer Gesellschaft among others) and the technical colleges form an integrated part of the industrial infrastructure of the Federal Republic. They cooperate closely with small and medium sized firms. The extended infrastructure in Germany facilitates more effective and widespread support. It can be expected that the three-fold increase in the Production Technology Programme budget in its second phase, for the 1984–7 period (more than DM500 million) will not be underspent or be limited in its effectiveness. It is significant for the priority attached to production automation in the Federal Republic that the budget for the programme represents a planned growth of more than 36 per cent per year, which is far more than the growth of any other R&D programme of the Federal Government.[29]

The countries which have adopted a support strategy of the machine tool industry aimed at military requirements (the USA, the UK and France) have not been very successful.[30] The disaster of the British machine tool industry developed in the 1970s. The French Government adopted a far more general and integrated strategy in its 'Filière Productique' programme which started under Giscard D'Estaing. As in other *filières* (production chains) the French aimed at national independence in all strategical parts of the chain. Under the socialist government the robotics programmes and rationalization measures of the machine tool industry of earlier date received wider attention and larger funds in order to make France one of the major forces in the coming international information society. The effectiveness of such programmes, however, is severely hampered by the fragmented French industrial infrastructure and the relatively weak

[28] Horn, Klodt and Saunders (1985), p. 80.
[29] BMFT (1984b), p. 89.
[30] BMFT (1984b), p. 81.

international competitive position of French electronics and machine tool firms. As a consequence the *filière* approach gradually altered in outlook. Instead of far-reaching self-reliance, promotion of cooperation with competent Japanese and American firms was stimulated. Foreign direct investment in some of the strategic parts of the chain was tolerated, and finally also selective imports were officially approved.

The history of MATRA is a case in point.[31] Next to RENAULT and CGE, the firm was considered to be one of the major 'poles' of excellence in the 'productique' strategy. The firm had three major automation subsidiaries: ROBOTRONICS in vision systems, SORMEL in assembly robots and DATAVISION in CAD. In 1985, ROBOTRONICS was sold to Alan–Bradley, and SORMEL for 49 per cent to ASEA. Only DATAVISION with its Euclyde System is successful, but the integration of several automation activities clearly cannot be reached any more within this pole.

As a consequence of the problems arising from stimulating production of highly complex systems, the French and the British Governments, especially, are forced to move from the stimulation of both production and diffusion towards mainly stimulating diffusion (as has been done rather successfully in France by the Fonds Industrielle de Modernisation). This has already been the main option for the smaller countries. The only other option is to look for more European cooperation (see Chapter Seven).

6.1.5. Subsidies for biotechnology

Since biotechnology, like the other core technologies, finds many applications, research in biotechnology is financed by many different agencies in any country. It is often not possible to isolate biotechnology relevant research because it can be an integrated part of, for example, an agricultural research programme (development of new seed varieties) or a programme on road transport (alternative fuels), etc. Therefore only a very rough comparison of expenditures in different countries is possible. The European Community has tried to make such a comparison (Table 6.4). The amounts have doubtless increased during the years following 1982/3.

European governments want to avoid a late start in biotechnology which would bring Europe into the same uneasy position that it actually has in the field of microelectronics. The figures presented in Table 6.4 show that at the beginning of the 1980s the European countries were behind the

---

[31] Interviews in Paris, June 1985.

Table 6.4   Estimates of public expenditures on biotechnology (in million ECU, 1982/3)

	Biotechnology 'Proper'	'Biotechnology-Relevant'
USA	225*	618*
Japan	56*	n.a.
West Germany	36	132
France	31	84
UK	46	59
Italy	13	34
Netherlands	10	26
Belgium	7	14
Denmark, Greece, Ireland, Lux.	3	6
European Community	146	355

Source: Commission of the European Communities (1983b), p. 34.
*At a rate of of $1 = ECU1.123.

United States in public support for biotechnology. In fact, the US Government spends more on biotechnology than the countries of the European Community and Japan together.

The first comprehensive public support programme for biotechnology in Europe was formulated in Germany in the mid 1970s. The German case demonstrates that cooperation between government, universities and business is often so close that it is difficult to trace who has taken the first initiative to subsidize specific activities. Business associations sometimes react to the request of government to draw up concrete proposals, but the idea to come up with such a request may have originated from the business association itself. The groundwork for a comprehensive federal policy in Germany was laid by a report prepared by the German Society for Chemical Engineering (DECHEMA), which effectively gave shape to the government programme. Most of the support went to the large chemical companies, first of all HOECHST, BAYER and BASF. The fact that DECHEMA had drawn up the first blueprint guaranteed that engineering problems ranked high on the list of priorities. But the government was very reluctant with regard to genetic engineering.

The federal support at first did not achieve the intended results. Early in the 1980s, a far-reaching reappraisal took place, sparked off by the highly publicized decision by HOECHST to fund basic research in molecular biology at the Massachusetts General Hospital in Boston,[32] because of the relative neglect of genetic engineering in continental Europe.

HOECHST had cooperated for years on one of the first biotechnology projects funded by the Federal Government: a project to develop a process

[32] Jasanoff (1985), p. 34.

for producing single-cell protein (SCP). The project remains one of the largest public–private undertakings in German biotechnology. Public support was, to a large extent, a response to the fact that a number of leading foreign companies, including Britain's ICI, were known to be experimenting with SCP processes.[33] The German Research and Technology Ministry had hoped for joint laboratories for the big companies and *German* research institutes. The decision of HOECHST to spend about $70 million on research in the United States provoked a change in the policy of the BMFT. It started to stimulate cooperation programmes between private companies and universities or other public research centres much more actively. This resulted in closer cooperation between, for example, BASF and the University of Heidelberg, between BAYER and the Max Planck Institute at Cologne and between SCHERING and the Technical University of Berlin.[34]

The early German biotechnology programme has influenced a number of other European countries (notably Great Britain, the Netherlands and France) to start similar programmes, but from the 1980s on, the impact of developments in the United States has been much stronger.

The British Government was already subsidizing the development of enzyme technology during the 1960s.[35] Massive government help has emerged relatively late.[36] The starting point of a more coordinated policy was the publication of the Spinks Report in spring 1980, named after Alfred Spinks, formerly Director of Research at ICI[37] and chairman of an advisory committee which suggested a considerable increase in government spending. One of the recommendations of the Spinks Report was the creation of CELLTECH, with half of its capital coming from the National Enterprise Board (NEB) and half from private institutional investors. In the meantime, however, NEB shareholding has been reduced.[38] The less than enthusiastic attitude of the Conservative Government with regard to state intervention has recently even led to plans to sell off one of the leading centres funded by the Agricultural and Food Research Council, the Plant Breeding Institute in Cambridge, to a private company like SHELL or ICI with major agrochemical interests.[39]

The government in 1983 borrowed three industrial experts (from ICI, GLAXO and BP) to monitor what British Industry was doing — or failing to do — in biotechnology. The group has identified several priority areas

[33] Jasanoff (1985), p. 30.
[34] *Research Policy*, **14**, 1985, pp. 23–8.
[35] Sasson (1983), p. 75.
[36] Farrands (1983), p. 37.
[37] Commission of the European Communities (1983b), p. 25.
[38] Yoxen (1985), p. 238.
[39] Yoxen (1985), p. 239.

where it believes Britain can and should be strong. The government is prepared to fund the cost — up to one third — of selected research and development projects.[40]

In France, industrialists' role in policy formulation has been less prominent than in West Germany or Britain. Public measures to support biotechnology were first discussed in the Pelissolo Report.[41] A 'mission des biotechnologies' was established the same year, which produced a planning document for biotechnology in France.[42] During the first 3 years (1983–5) of a 10-year mobilization programme, about FF600 million were to be invested.[43] These plans proved to be overambitious, however. According to the Office of Technology Assessment of the United States Congress, 'enthusiastic French Government officials advocated generalized support of R&D projects regardless of the prospects for successful exploitation, to the dismay of industrialists who doubted the viability of some of the projects designated to receive Government support'.[44]

Since the nationalization of major companies by the socialist government, it has become more difficult to distinguish between government policy and company strategy. Of the three major French companies with important research programmes in biotechnology, ELF–AQUITAINE is 67 per cent government owned, RHÔNE–POULENC 100 per cent , and ROUSSEL UCLAF 40 per cent (with the rest owned by HOECHST of West Germany).[45]

The Dutch Government decided in 1984 to subsidize GIST–BROCADES with DFL100 million although the firm had not explicitly requested such funding. In fact, the firm stated afterwards that it would have developed comparable expertise but perhaps somewhat more slowly. The government decision prompted a reaction from another Dutch company (DSM) to apply for comparable support. With a few large firms active in similar fields, selective support for individual firms leads to a chain reaction in subsidy schemes and to a quick rise of the total amount of subsidies.

In addition to the support from the different member countries, support for the development of biotechnology also comes from the European Community. As early as 1975, the Commission issued preparatory documents for a Community programme in biomolecular engineering.[46] It was more than six years, however, before the Council of the European Community finally adopted the Biomolecular Engineering Programme

[40] *Financial Times*, 11 May, 1984.
[41] Pelissolo (1980).
[42] OTA (1984a), p. 477.
[43] Sharp (1985a), p. 189.
[44] OTA (1984a), p. 479.
[45] OTA (1984a), p. 75.
[46] Commission of the European Communities (1983b), p. 15.

(1982–6). The reason for this delay is that the Community proposals had contributed to the awareness of governments of the potential of biotechnology. Governments wanted to devise their *national* programmes before embarking on a common programme. The endowment of the first Community programme, accordingly, was very meagre (ECU15 million in total).[47]

Private companies have not been very interested in the programme. Of the more than one hundred research contracts concluded during the first phase, about half have been with university departments and the other half with public research institutes. In only two projects, private companies participated (DEGUSSA and a small Greek company). German companies feared that the German contribution to a European programme would be deducted from the biotechnology funds of the BMFT, and that public support consequently would diminish rather than increase as a result of a European programme, because not all the money would flow back to Germany, while the red tape to get the funds would multiply. The French were not sure of a 'fair return' on their contribution either, given the lack of staff at the few institutions in France which would be able to submit applications. While governments of the larger member countries, therefore, have been very hesitant with regard to Community programmes, support came first of all from smaller countries with necessarily less comprehensive national support programmes.

The reluctance of large companies and the larger member countries has also been the reason for the important cuts that the proposals for a research action programme of the EC in the field of biotechnology for the years 1985–9 have undergone. Whereas the Commission had a total budget of ECU200 million (or ECU40 million a year) in mind, the Council of research ministers finally approved only ECU55 million, spread over five years (or ECU11 million a year). Quantitatively, the common European programme is not very impressive. It is less interesting for private industry than it is for universities. But the deliberations that lead to such a programme are important for the preparation of the *national* programmes in Europe. The preparation of international programmes leads to a continuous comparison between national programmes and draws the attention of national policy makers to important fields they may not yet have covered.

To increase government support for a European approach, the Commission has tried to follow an ESPRIT-like approach and mobilize the support of large companies that in turn could influence their national governments. The chief executive officers of sixteen European biotechnology companies were invited in December 1984 to discuss possible

---

[47] Commission of the European Communities (1983b), p. 9.

measures and eventual cooperation.[48] It turned out, however, that the interests of these companies are so diverse that it is much more difficult to devise a common programme than in microelectronics. Whereas microelectronics companies come at least from the same industrial sector, biotechnology companies have a much more diverse background (petrochemical industry, pharmaceutical industry, food industry). It is more difficult, therefore, to agree on common priorities and to start a network of cooperation at the European level (see also Chapter Seven).

6.2. GOVERNMENTS AND LARGE COMPANIES: A CHANGING BALANCE OF POWER?

Having described major government support programmes for some of the core technologies in the first part of this chapter, we now turn to a more detailed analysis of who the beneficiaries of these programmes are. In 6.2.1, we shall see how far national R&D subsidies concentrate on large enterprises. But government support not only takes the form of subsidy programmes. Governments increasingly try to use public procurement for the stimulation of innovation processes. Therefore, the prominent role played by many of the European high tech multinationals as prime defence contractors is looked at (6.2.2), and the bargaining process between the large telecommunications companies and the national PTTs is analysed (6.2.3)—with the objective of finding an answer to whether the balance of power between national governments and large companies has shifted as a result of the internationalization process and fast technological change.

6.2.1. Concentration of national R&D on large companies

Industrial R&D expenditures are largely dependent on a limited number of multinationals. In the United States and the United Kingdom the distribution is very unequal with 80 per cent of funds spent in large firms (employing more than a thousand workers) and less than 5 per cent in small firms,[49] These figures, however, do not specify the number of firms which make up

---

[48] Participants were: representatives of Rhône–Poulenc, Dechema (German Association of Chemical Machine Building), the European Federation of Biotechnology, Hoechst, Vyoril, ICI, Novo, Celltech, Montedison, Gist–Brocades, Akzo, Solvay, Wellcome, Elf–Acquitaine and Amylum.

[49] OECD—Science and Technology Indicators, Paris (1984).

Table 6.5   R&D Budgets of four–five major firms compared to total gross domestic expenditures on research and development (millions of ECU), 1975–83. (% of GERD)

	1975		1980		1983
GERMANY					
Siemens	528	Siemens	1 188	Siemens	1 542
Hoechst	305	Hoechst	515	Bayer	746
Bayer	263	Volksw.	495	Hoechst	712
Daimler	223	Bayer	492	Daimler	661
AEG	248	Daimler	436	VW	617
	1 567 (22%)		3 126 (22%)		4 278 (22%)
NETHERLANDS					
Philips	638	Philips	998	Philips	1 261
Akzo	100	Shell*	185	Shell*	306
Shell*	100	Unilever*	132	Akzo	195
Unilever*	67	Akzo	149	Unilever*	182
	905 (69%)		1 464 (61%)		1 944 (69%)
USA					
Gen. Mot.	898	Gen. Mot.	1 597	Gen. Mot.	2 924
IBM	762	Ford	1 203	IBM	2 825
Ford	602	IBM	1 092	AT&T	2 799
AT&T	499	AT&T	961	Ford	1 967
Gen. El.	288	Un. Tech.	474	Un. Tech.	1 091
	3 049[†] (11%)		5 327[†] (11%)		11 606[†] (12%)
JAPAN					
		Toyota	375 (est.)	Toyota	1 100
		Hitachi	356	Hitachi	1 057
		Matsush.	342	Matsush.	823
		Nissan	286	Toshiba	666
		Toshiba	222	Nissan	622
			1 581 (10%)		4 268 (13%)
FRANCE					
		Thomson	664	Thomson	679
		Renault	292	Renault	416
		CGE	278	CGE	408
		PSA	251	Rh.–Poul.	322
		Rh.–Poul.	233	Elf.–Aq.	311
			1 718 (19%)		2 136 (17%)

(continued)

Table 6.5 (continued)

	1980		1983	
SWITZERLAND				
	Ciba–G.	402	Ciba–G.	668
	BBC	386	Hoffm.-L.	532
	Hoffm.-L.	298	BBC	534
	Sandoz	178	Sandoz	294
		1 264 (72%)		2 028 (81%)
SWEDEN				
	Volvo	271	Volvo	368
	Ericsson	172	Ericsson	289
	Saab–Sc.	119	Saab–Sc.	238
	ASEA	110	ASEA	176
	Alfa–L.	35	Alfa–L.	46
		707 (40%)		1 117 (43%)
ITALY				
			Fiat	412
			STET	245
			Montedis.	193
			Olivetti	95
			ENI	58
				1 003 (21%)
UK				
			GEC	902
			ICI	470
			Plessey	368
			BP	346
			Shell*	306
				2 392 (20%)

Source: Wiltgen (1987).
*R&D figures of Shell and Unilever are taken into account for 50% in the Netherlands and for 50% in the UK. In all other cases the budgets are entirely related to the country of origin.
† For the American companies only company-sponsored R&D is taken into account. The budgets of the firms in other countries also contain subsidized and contract research.

for the bulk of R&D capability. The smaller this number is, the larger their impact on the national R&D capability and thus the greater the bargaining power can be expected to be. In Table 6.5 figures are compiled on the concentration of national R&D spending on the four or five firms considered to be (among) the biggest R&D spenders.

Most multinationals do some R&D abroad. Since part of the large companies' research expenditures are spent abroad, Table 6.5 somewhat overstates the share of these companies in the national R&D efforts.

The Japanese firms are the least international in their R&D expenditures and multinationals from the small countries are the most international, although firms like IBM and General Motors, for instance, also spend a considerable share of their R&D budget abroad. If we consider, for example, that PHILIPS spends 50 per cent of its budget outside Holland, the share of the four largest companies in Dutch R&D falls to 45 per cent in 1975, 40 per cent in 1980 and 47 per cent in 1983. With regard to the Swiss companies (spending around 25 per cent abroad) the ratio would fall to 54 and 61 per cent in 1980 and 1983 respectively. These ratios, however, remain far higher than the ratios in larger countries even after this correction. In general, therefore, the following pattern can be observed:

(a) The United States and Japan show a relatively low concentration level (10 to 15 per cent). Larger European countries have a slightly higher level of R&D concentration (17–25 per cent), whereas in smaller European countries with large 'national' multinationals like Sweden, Switzerland and the Netherlands, industrial R&D is dominated by a few multinational firms which account for more than 40 per cent of private R&D expenditures.

(b) Over the 1975–83 period the share of the largest firms in the national R&D capability has remained fairly stable. The fact that the growth of national R&D expenditures and that of the first four or five multinationals runs parallel might be an indication of the interdependence of both positions.

(c) If the share of these five major firms is related to the overall *industrial* R&D of a country only, their domination would be still higher. In the UK, for instance, the five largest R&D spenders accounted for 41 per cent of all industrial R&D in 1978.[50]

Whereas the R&D capability of small firms has been very low in the UK and the USA, the situation in Germany, France and Japan is markedly different in the sense that smaller firms still account for around 20 per cent of total industrial R&D.[51]

Governments wanting to boost national R&D efforts have heavily leaned on a limited number of firms, even if they have resorted to generic measures. It is, for instance, obvious that a general measure to abolish income tax payments for R&D personnel favours the large R&D spenders since small firms will hardly employ people for R&D tasks exclusively. Direct R&D subsidies to generate specific vital technologies are given to firms with strong laboratories and with good contacts with universities.

[50] Op.cit., p. 114.
[51] 1978 figures. Op.cit. p. 114.

These are, with the exception of a few 'high tech' start-up firms, the large established companies.

If a country like France has targeted robotics as a core economic sector, the government cannot do otherwise but rely on the research centre of RENAULT which alone accounts for about 50 per cent of the whole research capacity of the robotics industry in France.[52] Comparable concentration ratios can be found in other national robotics industries and are true also for telecommunications and data processing in all European countries.

The result of the interaction of state policies and a limited number of firms doing the bulk of the research is that the R&D budget of many large firms contains important public funding. Seventy-six per cent of all R&D with PLESSEY, for instance, is 'customer funded'.[53] The largest customers of PLESSEY are the Post Office and the Ministry of Defence. With some of the major French firms, public funds amount to around half of their R&D budgets. It has been estimated that even 90 per cent of R&D in the 1970s of CIT–Alcatel, CGE's telecommunications subsidiary, was publicly financed.[54] In the 1980–4 period the state-funded share of the R&D budget of THOMSON ranged between 40 and 51 per cent.[55] In the same period, the state R&D aid for BULL more than tripled and the government-supported share of the BULL budget rose from 13 to 24 per cent.[56] SIEMENS notes that only 5 per cent of its R&D budget comes from public funds.[57] Some other research, however, especially in telecommunications, is indirectly funded by the government which pays a price for the installations it purchases that is understood to recover a large share of R&D costs incurred. For specific research projects—the Megachip, for instance—the public share can exceed one third.

It is clear that in specific targeted sectors perceived by governments as being of national interest, the interdependence of large European multinationals and governments is particularly strong. This has always been the case in defence, aeronautics and telecommunications, but becomes increasingly true in the area of other core technologies such as integrated circuits, new materials and biotechnology as well. For an analysis of the bias of state support in favour of large companies, it is not sufficient to compare the percentages of R&D budgets that go to the largest companies.

[52] Jeandon and Zarader (1982), p. 119.
[53] McKinsey (1983), exhibit 46.
[54] Rhodes (1985), p. 7.
[55] Thomson, *Annual Reports*.
[56] Which is still far smaller than the state impact on Thomson's budget especially due to the latter's function in the defence base of the nation.
[57] Siemens, *Annual Report, 1985*.

These figures have to be seen in relation to the business structure of the country concerned. In the United States, 90 per cent of government R&D support goes to firms with more than 10 000 employees, and only around 8 per cent are channelled towards firms with less than 1000 employees, whereas in France, firms with over 5000 employees receive 'only' 80 per cent of total government R&D funding. However, the distribution of government R&D support in the United States is more or less a reflection of the business structure with 90 per cent of the funds going to firms which account for 84 per cent of total industrial R&D, while in France, those firms that receive 80 per cent of total government R&D funding contribute only 60 per cent to industry-financed R&D. More than half of the state subsidies for R&D in the 1970s went to six companies only: CGE, THOMSON (telecommunications and electronics), DASSAULT and SNIAS (armaments) and CREUSOT LOIRE (nuclear reactors).[58] The smaller firms received just 8 per cent of government funds (in 1981) while they accounted for around 20 per cent of total industrial R&D expenditures. Public support in France, therefore, discriminated much more in the early 1980s in favour of large companies than it did in the United States.[59] The figures for *research* subsidies for large French companies, however, may be inflated. An inquiry by the Court des Comptes in 1984 revealed that the aid distributed by the research directorate of the Ministry of Industry had been misappropriated and used as operating subsidies or to cover losses rather than to contribute to a coherent and offensive R&D strategy.[60]

The spending of the *German* government of R&D in electronics throughout the 1970s concentrated on three firms: SIEMENS received between 25 and 30 per cent of these funds, AEG between 10 and 15 per cent, while Valvo (PHILIPS's Hamburg subsidiary specializing in chips) on average received around 10 per cent of these funds. These three firms, therefore, accounted for more than half of all the funds provided for microelectronics research.[61] To give an impression of the amounts of money involved: in the 1973–82 period SIEMENS received DM4211 million (around half for nuclear energy R&D) and AEG DM480 million.[62] In general, in the beginning of the 1970s, 75 per cent of BMFT funding was in support of only eight large companies. However, the ministry

---

[58] *Outlook on Science Policy*, **1** (18), November 1979, p. 7.

[59] Direct government funding of R&D by size of firm percentages derived from: OECD (1986d), p. 272.

[60] Court des Comptes, La gestion des crédits 66–04 'Fonds de la recherche scientifique et de la technologie', Rapport au Président de la République, Paris, Journaux Officiels, pp. 39–46; cited in Rhodes (1985), p. 6.

[61] OTA (1983), p. 409.

[62] Webber *et al.* (1984).

[63] Horn (1982), p. 59.

notes in its 1979 research report that this has declined: six firms received 44 per cent of BMFT funds (and 24 firms 63 per cent).[63] Finally, in the 1970s the *United Kingdom* state R&D support was for 75 per cent aimed at five large companies, and 98.5 per cent was received by the 100 largest companies.[64]

The OECD concludes from comparable observations: 'High R&D intensity industries are responsible for about half of all R&D expenditures in manufacturing industry and get between 60–80 per cent of direct government R&D grants received by manufacturing industry.'[65] Rothwell and Zegveld come to the general conclusion that governments have lacked an active policy and merely followed the strategies of the dominant firms. With the result that

> the bulk of public R&D support has gone mainly to assist development activities in large companies. Most of these funds have been concentrated on large, prestigious projects in a few sectors of industry (aerospace, computers, defence, nuclear energy); of the remainder, a significant proportion has gone in support of marginal, high technical risk, relatively low market potential projects of the sort that the large companies would not themselves fund wholly out of their own resources. . . . In the case of private sector industrial support, in most countries loans and equity participation has also been concentrated in large companies. Thus, finance for the technological development activities and growth of small firms generally has been scarce during most of the post war era.[66]

Due to the extremely favourable climate for support in 'high technology' projects as a result of the international race, probably also many second or third best projects have been funded by government agencies not having the expertise to assess the value of the proposals. Proposals of smaller companies often have been scrutinized far more intensively than those of larger companies. This has led the Swedish and German governments to adopt a policy of mainly generic technology support. The Swedish Government argues that (large) firms would invest anyway in economically viable projects.[67]

Such a strategy, however, is only a realistic option for countries with a strong and technologically competitive industry and a specific industrial

---

[63] Horn (1982), p. 59.

[64] John Hagedoorn, Een verkenning van het innovatiebeleid in de lidstaten van de Europese Gemeenschap, *Tijdschrift voor Politieke Ekonomie*, **4** (4), June 1981.

[65] OECD (1986b); only Japan is an exception in 'that the relatively little direct government finance for manufacturing R&D — about 2 per cent of the total — is mainly destined for low R&D intensity industries (60 per cent)' (p. 13).

[66] Rothwell and Zegveld (1985).

[67] The Swedes were probably also inspired by the problems they had experienced in the 1970s with their effort as a small country to sustain a large and home-based mainframe computer capability beside Ericsson. See 6.1.4.

structure (as those in Sweden and Germany). Many factors come together that make it hardly possible in most European countries that a strategy of more support for smaller firms is really chosen. The very same factors would limit the effectiveness of such a strategy if nevertheless embarked upon.

(a) One of the major reasons for governments to subsidize R&D is that many research avenues are too expensive and too risky for any individual firm (e.g. fifth generation computers, megabit chips, new telecommunications switching equipment, etc.). Because of the large amounts of capital needed and the high development risk, small firms are almost by definition excluded from this kind of project.

(b) Most Western governments face budget deficits and follow an austerity policy which often entails a reduction in the number of civil servants. A reduced number of civil servants will be even less able to take the needs of small firms into account and to evaluate, process and follow up a large number of applications from small companies with relatively small amounts of money involved in any single case (see the example in 6.2.2).

(c) Many government departments have a long history of support for large firms and derive legitimacy from this intimate relationship. Propositions for subsidy schemes aimed at reaching small and medium sized companies, often meet considerable reservations inside government departments with vested interests in cooperation with large companies. These reservations sometimes can only be overcome if the large firms themselves demand support of smaller companies which serve as their subcontractors.

(d) Efforts to promote projects with a high commercial value are normally accompanied by the installation of commissions of experts from the business community to assist civil servants in the preparation of support programmes and in the evaluation of applications, since civil servants are regarded as being too far away from the market. Normally, these experts will come from large companies. Being less specialized than small companies, their representatives are expected to have a broader view. Besides, small companies can hardly afford to miss key personnel for the time needed for intensive participation in commission work.

(e) Large companies are in a better position to reserve scarce resources for 'subsidiologists' or lobby activities in general. Research with 27 large companies in the Netherlands, for instance, showed that at least 20 of them had a special department, commission or manager assigned to deal with public servants and politicians.[68]

---

[68] *De Volkskrant* (Amsterdam), 1 April 1986.

(f) Sometimes, government officials themselves come from large companies, or, more often, eventually hope to find a (higher paid) job with these firms. This is, for example, often the case in France where civil servants and managers of large companies share the same background of the *grandes écoles*, and appointments in public service belong to the typical career pattern of managers of large companies.

The intricate interrelationship between the public administration and the large companies makes it difficult to realize a shift of subsidies in favour of smaller companies. New political priorities are not enough. The whole institutional apparatus would have to be geared explicitly towards such a goal. The promise of large support for the creative potential of small and medium sized enterprises otherwise will remain purely rhetoric.

6.2.2. European multinationals as prime defence contractors

Throughout history, the development of new technologies has been closely linked to military efforts. This is true for the present core technologies as well. The technologies targeted by the military are largely the same as our core technologies (including new materials, but not — yet — putting much emphasis on biotechnology).[69] Cooperation between large companies and national governments has traditionally been very strong in the field of armament production. And while companies have become more and more multinational in almost all fields of activity, they have remained rather national with regard to military research and production. This is even true for a company like PHILIPS, one of the most internationalized companies in the world. It is assessed that at least 7 per cent of PHILIPS's total direct sales consists of military products. While only one fifth of all PHILIPS's employees work in the Netherlands, about half (6000) of these 13 000 employees involved in alleged military production are concentrated in PHILIPS's 'homeland'. Whereas the share of personnel engaged in production for the military is 3.7 per cent worldwide, it is about 8 per cent in the Netherlands (where direct sales to the military account for 10 per cent of total sales.)[70]

---

[69] The European Ministries of Defence, cooperating in the Independent European Programme Group (IEPG) in mid 1985, targeted five so-called Cooperative Technology projects. Two are in the field of microelectronics (gallium arsenid technology and image processing) and two are in the field of new materials (very low-weight materials and composites). The fifth project is on conventional ammunition. See Ministerie van Defensie — nota Defensietechnologie, The Hague, Tweede Kamer der Staten-Generaal, Vergaderjaar 1985–86, 19404, p. 11.

[70] Louisse (1983).

Table 6.6   *Percentage of government R&D expenditure on defence, 1980–4*

	1980	1981	1982	1983	1984
USA	47.3	51.8	56.9	n.a.	n.a.
Japan	2.3	n.a.	n.a.	n.a.	n.a.
FRG	10.0	8.8	8.5	8.4	9.8
France	36.5	37.2	35.1	29.7	31.3
UK	54.1	52.0	52.2	50.0	49.6
Italy	2.7	6.5	6.7	6.1	8.9
Netherlands	3.1	3.0	3.0	3.0	3.1
Sweden	15.6	16.0	19.2	n.a.	n.a.
Belgium	n.a.	n.a.	n.a.	0.4	0.5
Denmark	n.a.	n.a.	n.a.	0.2	0.2

Sources: OECD, Basic Statistical Series, Vol. A, *The Objectives of Government R&D Funding (1974–85)*, 1983, Paris; Directorate-General for Science, Research and Development, in *European Economy*, no. 25, September 1985, p. 106.

A very large share of government R&D subsidies is spent on defence-related projects (see Table 6.6). There are remarkable differences from country to country as far as the share of government R&D support devoted to military projects is concerned. This share ranges from more than half in the case of Great Britain (where it exceeded even the share in the United States in the early 1980s) to some 3 per cent in the Netherlands and even less in other small European countries (see Table 6.6). The relevance of military contracts for large companies not only lies in the fact that they account for a large share of government R&D support. They also provide a pattern of interaction between government and industry which is followed in civil sectors, where governments have subsidized their national champions in electronics as if they operated on a cost-plus basis in civilian industry as well.

Great Britain and France (like the United States) spend a very high share (10–12 per cent) of their total military budget on R&D. Their defence budget shows a much stronger commitment to R&D than the budget of other countries. Apart from Sweden (with 7 per cent of the defence budget allocated to R&D) the other European countries have shares of 3 per cent or below.[71] The interaction between R&D policies of state bureaucracies and the defence industry in the first four countries can be expected to be far more intense than in the latter countries. Germany and the other low-profile defence spenders rely more on the research and development done

[71] Germany 3 per cent and the Netherlands, Japan, Switzerland and Italy between 1 and 2 per cent. See Ministerie van Defensie—nota Defensietechnologie, The Hague, Tweede Kamer der Staten-Generaal, Vergaderjaar 1985–86, 19404, p. 25.

by the industry itself and therefore can only influence the strategy of multinationals in the area of defence technologies by way of their procurement policies.

The high priority for R&D spending on military technologies in the UK and France has influenced the overall industrial capabilities of major firms. 'Large public expenditures on defence-related R&D, and indeed on defence spending generally, appear to carry with them high opportunity costs associated with the diversion of resources away from developments associated with economic growth, productivity increase and industrial competitiveness in civilian markets.'[72] As a consequence there is an inverse relationship between economic and productivity growth rates and rates of military expenditure in general.

In the UK, for instance, major companies (such as PLESSEY, GEC, FERRANTI) in the 1970s decided to concentrate their efforts on military requirements.[73] This means high-performance, low-quantity production. In order to be able to concentrate on this market, GEC for instance, voluntarily chose to withdraw from standard chip production (lower reliability needed, but far larger quantities). MATRA–HARRIS SEMICONDUCTORS has also adopted a niche strategy in the production of chips especially designed for military purposes.[74] In commercial markets, however, this puts the British and the French industries in a very disadvantageous position. British and French companies seem to have followed the American example, where a more and more specialized defence industry has come into existence with fewer and fewer links with civil production. In the other European countries, civil and military production is more often carried out by the same companies, with civil and military departments sharing the same technology base. In Germany, DAIMLER–BENZ has become the prime contractor of the Defence Ministry, after having taken over MTU, DORNIER and AEG (with SIEMENS being third and KRUPP the seventh on the list of defence contractors).[75] DAIMLER–BENZ thus followed the example of GENERAL MOTORS in the USA which became the prime contractor of the Pentagon after it had taken over HUGHES AIRCRAFT and EDS.[76]

The concentration of public support on the large companies is accentuated at a time when ministries have come under pressure to reduce

---

[72] Rothwell and Zegveld (1985), Chapter V.

[73] McLean and Rowland (1985), p. 19. In 1976, for example, the Ministry of Defence funded 54 per cent of electronics R&D, and in the late 1970s, the British ministry was taking up over 20 per cent of the electronic industry's output, which implies that the total share of military sales (i.e. including export in the output of the British electronics industry was at 30 per cent. See Sami Faltas, Warships and the world market (1985), dissertation, Amsterdam, p. 253.

[74] The strength of a niche strategy, *MATRA Newsletter*, 1985.

[75] See *Der Spiegel*, no. 43, 1985, p. 144.

[76] See *Business Week*, 27 January 1986, p. 64.

*Table 6.7   Share of telecommunications service providers' purchases
in domestic sales of telecommunications equipment (1980)*

	%		%
USA	73	Japan	32
W. Germany	80	France	66
UK	76	Italy	85
Sweden	71	Netherlands	70
Belgium	65		

Source: OECD (1983a), p. 131.

their staff. This is explicitly acknowledged in a booklet *Selling to the MOD*
by the British Ministry of Defence which states that 'in a drive to reduce
costs and staff numbers in the MOD, and in the interest of efficient project
management, we are seeking to place more and more of our defence
contracts in the hands of a single prime contractor'. The MoD urges small
subcontractors to address themselves not to the MoD but to the defence
primes.[77] This pattern supports the growing importance of core firms in
a subcontracting system as analysed in Chapter Three.

6.2.3. The dominance of national champions in the telecommunications industry

This concentration is most obvious in the case of telecommunications
equipment. The role of public procurement is almost as strong as in the
defence industry (see Table 6.7). Besides, both industries overlap to a
considerable extent, with a considerable share of all telecommunications
equipment being sold to the military.

The relationship between the prominent national suppliers of telecommunications equipment and the national telecommunications service
(which until recently has been a publicly monopoly in practically all
European countries) has been very close since the inception of the industry.
The recent introduction of digitalization has led to a certain reshuffling
among the important suppliers, but in general, the position of the 'national
champions' in their respective national markets has been strengthened

---

[77] In the words of the ministry itself: 'Given this devolution of power to the defence
primes and the non-interventionist policy of the MOD, it is the purchasing policies of the
primes which control the distribution of defence subcontracts. The volume of work available
to small firms (within a given defence budget) is not regulated by MOD and is determined
by the amount of work which the defence primes are willing to contract out.' *Selling to
the MOD*, a booklet to help small firms and component manufacturers to compete for MoD
business, undated, p. 8.

rather than weakened (see Table 6.8). An exception to this trend is the situation in the UK where British telecommunications companies (GEC and PLESSEY) meet increasing difficulties in competing even in the domestic market with their digital exchange equipment (System X), which does not sell well in the international market either.[78]

Whether public telecommunications services can take a relatively independent stance with regard to their major equipment suppliers or become mere captives of the major companies depends to a large degree on the extent to which they maintain an independent research and development capacity which allows them to evaluate and compare the offers of different companies and to specify their own standards instead of having them formulated by the companies. The difference in this respect between France and Great Britain on the one hand Germany on the other is striking. One of the main reasons why the purchasing agency in France did not become dominated by the supplier was the development of an independent technological capacity via CNET (Centre National d'Études des Télécommunications), which was indispensable to the Direction Generale de Télécommunication (DGT) in organizing the market.

> In Britain, first the Post Office then British Telecom had a strong research capability in the Martlesham centre which in the case of the development of the new generation of exchanges — System X — gave BT a good deal of autonomy in its negotiations with suppliers. The system was specified within a close bargained network through a committee on which the supplier firms were represented, and the division of labour between BT, Plessey and GEC was the outcome of negotiations. In Germany, the Bundespost relied more heavily on the technological expertise of the dominant supplier, Siemens, and technical specifications written into its tender invitations reflected this.[79]

The situation in Sweden is similar to the one in France and Britain. Televerket is one of the largest state authorities, with an own manufacturing company, TELI, formed in 1981. Televerket in 1970 formed a joint

[78] A more ambivalent situation has come about in West Germany: although Siemens did increase its own share in public purchases of telecommunications equipment, it lost some of its dominance since the second supplier, formerly ITT-owned Standard Elektrik Lorenz (SEL), no longer produces on the basis of a Siemens licence, but offers its own 'System 12', developed in the European laboratories of ITT.

[79] Cawson, Holmes and Stevens (1985), pp. 36–7. They continue: 'The supply of the older generation of electro-mechanical switching equipment up to 1974 was confined to four firms, but three of these were producing under licence from Siemens. It was a consortium of these four firms under the leadership of Siemens which was developing a new computer-controlled (but analogue) system for the Bundespost when the latter came to the conclusion that its decision had been mistaken, and that digital exchanges would conquer the world market. But even here it was the firms who pressed this argument on the Bundespost, which maintains a stance of not attempting to influence company strategies.'

Table 6.8   *Public switching equipment market in Europe, 1974–87*

	1974	1985–7 (digital)
Austria	SIEMENS ITT local (ITT patent)	SIEMENS ITT 50% local firm 50% (Northern Tel. patent)
Belgium	ITT 80% GTE 20%	ITT(CGE) 66.6% ATEA–SIEMENS 33.3%
Denmark	ERICSSON 70% ITT 10% SIEMENS 20%	ERICSSON 80% ITT 20%
Finland	ERICSSON 60% ITT 15% SIEMENS 25%	Telenokia (CGE patent) 50% ERICSSON 30% SIEMENS 15%
Spain	ITT 25% ERICSSON 25%	ITT 70% ERICSSON 30%
France	ITT 42% ERICSSON 18% CIT–ALCATEL 40%	CIT–ALCATEL/THOMSON 84% CGCT 16%
UK	PLESSEY ITT GEC	PLESSEY/GEC THORN–ERICSSON APT
Greece	SIEMENS 40% ITT 40% PHILIPS 15%	ITT n.a.
Ireland	ERICSSON 65% ITT 35%	CIT–ALCATEL 40% ERICSSON 40%
Italy	ITALTEL 50% ITT 20% GTE 5% ERICSSON 15%	ITALTEL/TELETRA/GTE 65% ITT ERICSSON
Norway	ERICSSON 60% ITT 40%	ITT 100%
Netherlands	PHILIPS 75% ERICSSON 25%	PHILIPS/ATT 60–75% ERICSSON 15–25%
Germany	SIEMENS 55% ITT (Siemens licence) 30% TEKADE 15%	SIEMENS 60% ITT 40%
Portugal	ITT 50% PLESSEY 50%	SIEMENS 50% STANDARD ELECTRA (ITT) 50%
Sweden	ERICSSON 100%	ERICSSON 100%
Switzerland	ITT 35% SIEMENS 30%	ITT SIEMENS

Sources: Dang Nguyen (1985); Labarrère (1985), p. 196; own observations.

company with ERICSSON. This company—Ellemtel Development Company (Ellemtel Utvecklings AB)—has been developing the successful AXE-System and other technologies. Richardson comes to the conclusion that

> Televerket has been an enlightened and progressive telecommunications administration in large part it is vertically integrated, i.e. it has developed its own R&D and manufacturing expertise. This has meant that Televerket has been able to devise a very effective purchasing policy, to the benefit of itself and of Ericsson. In selling to Televerket, Ericsson has faced a purchaser having equal expertise and always willing to purchase overseas if necessary.[80]

But in most smaller countries—and more and more in the larger countries as well—the PTTs are becoming increasingly dependent on the technological possibilities of the large telecommunications producers. They probably will mainly *adapt* the standardized equipment of a few suppliers to the national market. In the negotiations with the large telecommunications firms the PTTs seem to have problems in getting highly trained personnel and thus have more limited power in bargaining towards their own specifications (which they are less and less able to formulate independently).

6.2.4. A shift in the balance of power between government and companies

In the preceding pages, different forms of public support for large companies have been reviewed: the concentration of R&D subsidies on large companies, the role of European high tech multinationals as prime defence contractors, and the symbiosis of public telecommunications services and the national champions in the telecommunications industry. If these different contributions to the cash-flow of Europe on high tech multinationals are taken together (see Table 6.9), the impressive dependence of some of them on national government spending becomes obvious. This is very much the case for some of the largest electronics groups, especially for those involved in telecommunications. This is less the case for computer producers, and much less for the automobile and (petro)chemical companies. The very close relationship between some of these companies and the national government of their home country makes it questionable to regard these companies as 'real' multinationals.

---

[80] Richardson (1986), p. 10.

Table 6.9 Assessing the closeness of the connection between multinational companies and European national states in the 1980s as percentage of their total turnover[1]

Firm	Defence contracts	Telecom sales	Importance of other public procurement	Direct subsidies	Home-based prod. (exp. %)[11] (1981)	State owned (1986)	Estimated share of turnover dependent on governments	
							World wide	Home state
TYPE A CONNECTION (>70% OF TURNOVER)								
STET	L352 billion (1982) (4.4%)	75% (1984)[9]	Dataprocessing subsidiary	Negl.	>90% (12%)	Yes	>80%	>70%
BRITISH TELECOM	n.a.	100%	Negl.	No	90–100%	No	>90%	>85%
TYPE B CONNECTION (20–40% OF TURNOVER)								
PLESSEY	22%	42%	n.a.	75% of R&D 'customer funded'	n.a.	No	>60%	>30%
MATRA GROUP	>30%[12] (1984)	8%	Space (10%)	FF688 m in 1982; 50% R&D budget in 1985	90–100% (50%) 1984	Yes 51%	50–60%	30–40%
THOMSON	>50%	Negl.[5]	Low (mainly cons. el.)	FF500 m (1984) 47% of R&D budget (1984)	86% (32%)	Yes	>60%	30–40%
SIEMENS	4–10%	30%	Consider.: (nuclear) power plants is 23% of turnover +8% medical technology	5% of R&D budget	85% (41%)	No	>60%	20–30%

GEC	15% (1977)	20%	Power plants (12% in 1985) Medic. equipment (9%)	Cons.	67%[3] (22%)	No	>50%	20–30%
STC/ICL	?	70% of STC	Cons.: ICL sales	?	?	No	>50%	20–40%
CGE	Low	>45%[7]	Railway and power plants (Alsthom–Atlantique 30% in 1984)	FF1861 m (1984) 30% of R&D budget (1984) Group Alcatel	60%	Yes	>60%	20–30%
TYPE C CONNECTION (10–20% OF TURNOVER)								
ERICSSON	6% (1984)	77%	Low	Low	56% (36%)	No	>80%	10–20%
BULL	n.a.	Low	Preferred procurement of computers	FF1 b (1985) 24% of R&D (1984)	>63% (1984)	Yes	10–40%	10–20%
THORN–EMI	10%[2]	5–10%[13]	Lower than 10% 10% (important cons. el. sales)	Low	72%[3] (5%)	No	15–20%	10–15%
DAIML./AEG	DM3699 m at least in 1983 (= 5–10% of total turnover)	7% of AEG	AEG: energy tech. (DM1390 m) Bahntechnik (DM301 m) (1984) Daimler: city buses	DM585 m of BMFT in 1972–82 period	80% (42%, 27% AEG)	30%[7]	15–25%	10–20%
ASEA	n.a.	Low	Power plants and related activities (22.5%) Transport (<6%)	Negl.	>39%	No	30–50%	10–20%
TYPE D CONNECTION (0–10% OF TURNOVER)								
FIAT	5%[2] (1982)	2% (Teletra, 1984)	Railway equipment equipment (1%)	Medium (restructuring)	(17%)[2] 1984	No	<15%	5–10%
KRUPP	5% (DM828 m in 1983)	Negl.	Power st. (1%), fire brigades (2%), rail, etc. (5%)	DM211 m in 1972–82 period	91% (37%)	No	10–15%	5–10%

(continued over)

Table 6.9 (continued)

Firm	Defence contracts	Telecom sales	Importance of other public procurement	Direct subsidies	Home-based prod. (exp. %) (1981)[11]	State owned (1986)	Estimated share of turnover dependent on governments	
							World wide	Home state
VOLVO	5–10%[2,15]	Negl.	Low	Negl.	55%	No	5–15%	5–10%
NIXDORF	n.a.	Low, but growing[14]	Low	3.1% (DM41 m) in 1981–5 of R&D	>49% (1985)	No	<10%	0–5%
PHILIPS	7%[2]	8%	Lower than 10% (traffic control and medical systems)	Consid. in specific areas	32% (22%)	No	15–25%	0–5%
RENAULT	Equipment division (1%)	Negl.	Low	FF3 b (1985)[4]	55%	Yes	<10%	0–5%
BOSCH	n.a.	26% (communications)	Low (power generators)	1% of R&D in 1984 (DM7 m)	74% (30%)	No	10–20%	0–5%
VW	Low	Negl.	Low	DM119 m from BMFT in 1972–82 period	67% (39%)	40%[10]	<10%	0–5%

OLIVETTI	Low	3.3% (1984) 5.3% (1983)	Computer sales to public agencies	Low	53% (20%)	No	<10%	0–5%
Mannesmann	n.a.	Interest in ANT	Low	DM161 m in 1972–82	69%	No	5–10%	0–5%
Rhône-Poulenc				FF400 m (1983) FF150 m (1984)	68% (29%)	Yes		0–5%
ST-GOBAIN				FF250 m (1983) FF150 m (1984)	54% (15%)	Yes		0–5%
BROWN BOVERI			Power plants	DM1380 m (1972–82) in FRG	15% (0%)	No		0–5%

[1] Not included in this assessment is the role played by most national governments as intermediary in the provision of medicines, which makes especially the pharmaceutical companies sensitive to government regulations, procurement and the like. Nor are the agricultural policies of countries included with regard to the food companies.

[2] Estimate.

[3] 1982.

[4] July 1986, the European Commission blocked this French support operation.

[5] After restructuring with CGE.

[6] With takeover of ITT's telecommunications division, the share will increase even more.

[7] Deutsche Bank.

[8] Thomson Brand.

[9] Like British Telecom, the STET group includes the major service provider. This is SIP. Together with the equipment manufacturer Italtel (9% of turnover), Italcable and Telespazio, these subsidiaries make up 75% of the total sales of the STET group.

[10] 20% Federal Government, 20% Lower Saxony.

[11] Both as percentage of *total* sales; *source*: Stopford and Dunning (1983).

[12] Estimate. More than 70% of turnover of Société Matra in 1984 was on military activities.

[13] Joint venture with Ericsson (Thorn–Ericsson).

[14] Nixdorf expects that the telecommunications business will become one third of total sales by 1990.

[15] Volvo is explicitly trying to replace a diminishing military workload in, for instance, its aerospace division with commercial products. In 1985 it succeeded in having larger commercial than military sales in this sector.

What does the large share of revenues from public funds mean for the government–company relationship? Does it provide governments with effective means to influence the decision making of companies? Or is the large share, on the contrary, more an expression of the power of these companies to influence public decision making in such a way that large amounts of money were allocated in their favour? It is probably not appropriate to draw any general conclusion for all the European companies mentioned, but it seems justified to state that, in general, recent developments have led to some shift in the balance of power between government and companies in favour of the latter.

With increasing technological sophistication and a shortening of technology life-cycles, government authorities have become more and more dependent on private companies to formulate their own innovation policies (if they do not restrict themselves to generic measures only). This is also true at the European level, where the twelve largest electronics companies have contributed in a decisive way to formulating the first large-scale cooperation programme (ESPRIT).

This shift in the balance in favour of private companies is also true with regard to ownership. It has to be kept in mind that about a quarter of the companies of our sample (see Chapter Two) are publicly owned. But public ownership gives public authorities less leverage than ever before. Fast technological development and stiff international competition make it increasingly difficult for national governments to prescribe any specific policy or strategy to management. The pressure to denationalize public companies, therefore, in most cases does not originate from the companies in question, but comes from the political arena. Politicians often want to sell companies, not only for ideological or budgetary reasons, but also to get rid of the political responsibility for the effects of the restructuring process (plant closures, loss of jobs).

The large-scale nationalization of French companies in the early 1980s at first seems to contradict this statement. Nationalization in France raised government control of the French industry in 1982 from 17.2 to 29.4 per cent of total sales. The number of employees in government service increased from 11.0 to 22.2 per cent, and the control of investments from 43.5 to 51.1 per cent.[81] Even more dramatic were the changes in the electronics industry. The nationalization of the major French electronics firms (BULL, CGE, MATRA (51 per cent), SAINT-GOBAIN, THOMSON) brought half of all the French electronics industry's production and well over half of all R&D in electronics under state control.[82]

---

[81] Savary (1984), p. 163.
[82] Webber, Moon and Richardson (1984), p. 39.

Judged on the basis of the *effect* of the nationalization, however, they first of all proved to be an effective means to channel even more public money towards the major companies, a transfer which otherwise would have been difficult to legitimize. It also made a new division of labour between the major companies possible and put an end to competing efforts in the same field by different companies. But it did not give government agencies a decisive say in company affairs, not even in the case of the highly loss-making company THOMSON which needed billions of government funds to cover its losses and depends for around 30–40 per cent of its sales on French public procurement (Table 6.9). This was not only a question of the personalities involved, but seems to be built into the French policy-making system. The 'Grand Corps' (of alumni of the elite institutions of higher education) still command great influence, and officials of public authorities, who themselves often aim at a career in a public enterprise, face great difficulties in 'controlling' senior management of nationalized companies, who are their 'superiors' in the hierarchy of the 'Grand Corps' and can have a decisive influence on their professional future (so-called 'pantoufflage').[83]

The dominant tendency actually, however, is not nationalization but *privatization*. Where privatization has not become an issue (for instance, with STET or DSM) this is often because the companies in question are run like private companies anyway. We shall elaborate on this problem in the next section by describing how the acceleration of technological development and the resulting pressure to internationalize have strengthened the tendency towards privatization and deregulation. Even where no changes in the legal status of the enterprises or in the laws that regulate their activities take place, the dominant tendency has an impact on the way the existing regulation is applied.

6.3. THE REGULATORY FRAMEWORK

Governments do not only support national companies by direct subsidies or procurement. Their support is much broader, embracing all the different forms in which governments influence the environment in which companies work—be it the labour market, the material infrastructure or the regulatory framework. State activity in this area has often been more important to companies than outright subsidies. The combined impact of the internationalization process and the development of new technologies has

---

[83] Cawson, Holmes and Stevens (1985).

led to important changes in this field as well. Especially the recent changes in the regulatory framework have changed the international scene for high technology enterprises. Two developments will be discussed in this section: the watering down of antitrust legislation and the general trend towards deregulation, particularly in telecommunications.

National antitrust legislation has increasingly come under pressure as a result of the internationalization process. The more national markets become internationalized, the less it seems justified to take market shares in the domestic market as the basis for antitrust regulations. Recent technological developments have put another time bomb under antitrust legislation. With the blurring of boundaries between different economic sectors caused by the development of new basic technologies (see Chapter One), it becomes increasingly difficult to delineate relevant markets. Companies from neighbouring branches that have experience in the same basic technologies become at least potential competitors in other branches as well. In spite of the fast pace of the concentration process, therefore, competition in many areas seems to have increased rather than diminished. The need for strict antitrust measures accordingly is felt less. In fact, the restrictions on mergers and intercompany cooperation imposed by antitrust legislation are more and more regarded as obstacles on the way to improving the competitiveness of national companies. National governments as well as the European Community, therefore, have taken steps to relax antitrust legislation in order to secure a better position for domestic companies in the international restructuring race.

The same trend can be observed for other regulations with regard to the environment, working conditions or state monopolies. Wherever these regulations are perceived to hinder experiments with new technologies, considerable pressure is exerted to lift these barriers in order to allow the national industry to enter new profitable fields. Since many governments — especially of smaller countries — lack the financial muscle to match the direct subsidies given by other governments to their national companies, they try to attract investment in high technology by offering an especially liberal regulatory framework. In addition to a subsidy race, countries thus have become engaged in a kind of *deregulation* race.

6.3.1. Anti-antitrust measures

The watering down of antitrust regulations that can be observed in most Western countries is a direct complement of the subsidy race in at least two respects.

On the one hand, governments hesitate to finance the same developments in different firms. If they do, they make a rather uneconomic use of scarce resources. If they do not, and concentrate on one firm instead, they can be accused of discriminating against the firm's competitors. The obvious way out of this dilemma is to let the firms cooperate. Cooperation in research often leads to cooperation in commercialization as well. (Otherwise cooperation in research becomes hampered the nearer the commercialization phase comes.) Where antitrust legislation stays in the way of such cooperation, it has to give way in most cases in the interests of international competitiveness.

The other link between the subsidy race and anti-antitrust measures is created by budgetary pressures. The departments responsible for support of research and development often not only hope that cooperation and concentration would make the national champions more competitive, but also that larger and more integrated companies would need fewer government subsidies in the future, thus aleviating the budgetary pressure in the medium run.

The restructuring race has thus led to more government-stimulated cooperation and even outright cartelization. Many governments tend to follow the *Japanese example* with its rather weak antimonopoly laws and legalized cartels. In the *United States*, the Reagan administration issued new rules governing mergers in 1982. Among the changes were a more liberal interpretation of market concentration and a greater willingness to approve even worrisome combinations if there were no major barriers to a new competitor who might want to enter the market.[84] The authorities also became more tolerant of cooperative R&D efforts. These joint projects, however, are only tolerated for American owned companies.[85] In February 1986, far-reaching proposals were presented to the US Congress by the administration for formal flexibilization of antitrust laws. Measures were proposed explicitly in order to consolidate the international competitive position of American companies.[86]

The Department of Trade and Industry in the *United Kingdom* determined in 1983 that as few mergers as possible should be subjected to a formal investigation by the Monopolies and Mergers Commission (MMC). In this, the department is in direct conflict with the MMC, but has the support of the government and the multinationals engaged in core technology areas. The result is, for instance, that BRITISH TELECOM was allowed to buy the Canadian telecommunications firm MITEL despite objections from the MMC.[87]

---

[84] *Fortune*, 1 September 1986, p. 49.
[85] So Mostek of United Technologies had to withdraw from one of the most prominent groups (MCC) after it was taken over by Thomson.
[86] *De Volkskrant* (Amsterdam), 21 February 1986.
[87] *Financial Times*, 28 January 1986. Other important recent takeovers in the IT sectors were those of ICL of STC and INMOS by Thorn–EMI.

France has never had a restrictive antitrust legislation. The French state historically has been an agent of concentration rather than an agent fighting concentration, which can be explained by the rather dispersed character of large parts of French industry.

The only major country with still comparatively stringent antitrust regulations seems to be *West Germany*. During the past years, a considerable number of cases of cooperation in the IT sector were not approved by the Bundeskartellamt. Examples are the proposed joint optical fibre factory by SIEMENS, PHILIPS, SEL, KABELMETAL and AEG in Berlin and the planned takeover of GRUNDIG by THOMSON. The Kartellamt is still influential in formalizing modes in which mergers, joint ventures or other forms of cooperation come about. It was not able, however, to stop the recent wave of large mergers in the German industry, the major move being the takeover of DORNIER, MTU and eventually AEG by DAIMLER–BENZ. As a consequence, the Bundeskartellamt becomes increasingly uneasy, while the interests of the Ministry of Economic Affairs and of the BMFT prevail.[88]

An erosion of antitrust regulation is also taking place at the level of the *European Community*. The EC Treaty in principle opposes interfirm cooperation. Until 1977 even pre-competitive cooperation in research was formally illegal. With the beginning of pre-competitive Community projects, this had to change. In 1985 a number of exemption categories were granted for extensive cooperation in *production* as a follow-up to the R&D cooperation programmes.[89] With a few exceptions, the directorate general responsible, however, has not been willing to extend this liberalization to the *commercialization* phase. But during the bargaining process around the EUREKA initiative, which aims at cooperation in commercialization as well, the European Commission has given up this opposition.

An obvious dilemma which accompanies the relaxation of antitrust regulations is that the increased flexibility provided might in the long run lead to structural economic rigidity, as exemplified by the French efforts to create national champions by stimulating concentration. Within a period of about ten years, the number of suppliers in the telecommunications equipment industry (supplying public digital exchanges) in France has been reduced from five to only one (ALCATEL–THOMSON). In other sectors (integrated circuits, robotics, etc.) a comparable process has taken place. In the case of telecommunications, this has been strongly opposed by the French PTT which prefers to deal with more than one supplier to make sure that the most advanced technology remains available and that

---

[88] See Interview with Bundeskartellamt-president (Wolfgang Kartte), Nach geltendem Recht sind wir machtlos, *Der Spiegel*, no. 47, 1985, pp. 114–27.

[89] See P. E. de Hen, De Mythe van Europa Inc. (3), *Financieel Economisch Magazine*, 5 October 1985.

prices are competitive. This preference resulted in a—still limited—opening of the market to a foreign supplier.[90] The very policy that aimed at mobilizing and concentrating all national resources in order to remain independent thus ironically finally implies an opening to foreign suppliers.[91]

6.3.2. Controlled deregulation

Anti-antitrust measures that allow for a higher degree of concentration have been complemented by other measures to ensure continuing competition. The watering down of antitrust regulations to reduce competition (at the national level) and deregulation to increase competition thus are not necessarily contradictory steps, but might complement each other.

National regulations often function as non-tariff barriers to protect national industry. Technical, health, security or environmental standards can be used to block the access of foreign competitors to national markets. In a period of very fast technological change, however, many of these protective measures can become counterproductive. They can slow down technological development by domestic firms and reduce their competitiveness in the long run. They may not only keep foreign competitors away from the national market, but also impede national firms wishing to expand abroad. Expansion abroad, however, can be necessary in order to reach the level of sales which would be needed to finance up-to-date research during the short period that demand for a new product is booming. State activities to support national industry at present may thus take the opposite direction (deregulation instead of regulation) from the 1970s.

In the context of support of high technology industries, the virtues of regulation and deregulation have been discussed most extensively for

[90] Throughout the whole of 1986 and a part of 1987 a fierce battle evolved over who was going to be allowed by the French Government to (partly) take over CGCT and its 16 per cent share of the switching equipment market. AT&T/Philips and Siemens were the most prominent contenders. For them the American and German Governments put in considerable lobbying efforts to open up the French markets for 'their' firms. Finally the French Government decided to allow a third producer (Ericsson in cooperation with Matra) the market entrance. This clearly was perceived as circumventing the political struggle which would ensue with either government, each being prepared to take considerable retaliatory measures against the French if it should be on the losing side in the deal. It is not to say, however, that this will not happen with the solution taken by the Chirac Government. In this context it is clear, though, that the small-country background (in which market the French industry had no real interest) in this case proved to be an asset for Ericsson.
[91] Which also seems to be the case of the UK defence industry.

telecommunications. Deregulation here can mean two different things, which are closely related, however.

(a) Deregulation can mean limiting or even abolishing the monopoly of the telecommunications carrier and eventually privatizing telecommunications services.
(b) Less far reaching, deregulation (in the sense of liberalization) can also mean that the concentration of procurement on a few established suppliers of telecommunications equipment is reduced, giving other companies a better chance to sell alternative equipment. Both aspects are closely related, because more competition for the telecommunications carrier(s) would increase the incentive to make use of the most sophisticated equipment whatever the source of supply. Restrictions of the (public) monopoly, therefore, would automatically imply an opening for new equipment suppliers.

The delicate balance between regulation and deregulation of telecommunications has been a hot item of discussion in all highly industrialized countries. Depending on the international competitiveness of the national champions, on the performance of the existing telecommunications network and on dominant government ideology, different compromises between the forces in favour of regulation and deregulation have been found in different countries. At least five groups of actors play an important role in the discussion of deregulation of the telecommunications sector.

(1) *Traditional prime contractors* to the national PTTs. Most European PTTs have one large prime contractor for public switching exchanges, which is also a 'national champion' and delivers more than half of the equipment (see Table 6.8).
(2) *Traditional prime contractors in other countries.* They look for economies of scale and try to become part of the privileged group of contractors in other countries as well.
(3) *Non-traditional contractors from neighbouring sectors* (data processing, television networks, cable producers, press agencies, publishers). These companies increasingly offer terminals, transmission equipment and/or specialized services (value added networks) that may compete with deliveries of traditional suppliers. These companies have a strong interest in the installation of the most advanced switching equipment, which would allow large-scale use of their products.
(4) *Large telecommunications network users.* Globally active multinational firms are interested in on-line communication between their plants and offices and opt for a free and uncontrolled flow of (digitalized) data. By expanding their private networks (offering it to outside users as well) they gradually erode the public telecommunications monopoly.

(5) *The PTTs themselves.* The public service (especially the trade unions organizing the employees) in general is in favour of a continuation of the public monopoly, arguing that any form of privatization would imply a reduction of employment. The higher echelons of management and technicians may not always agree with this stand, arguing that only a competitive organization would be able to attract and pay adequately the specialists needed for fast modernization.

The constituents of categories 4 and 5 are clear. In Table 6.10 major members of the first three groups are listed.

The firms in categories 3 and 4 are the clearest protagonists of deregulation, whereas the firms in group 2 are trying to take over the position of one of the other traditional prime suppliers without pressure for deregulation, since a preserved monopoly would protect their interests as well. If firms from the second group, however, do not succeed in breaking into the intimate group of suppliers of other national PTTs, they might team up with firms in categories 3 and 4. Their activities abroad, however, may contribute to a chain reaction of deregulation, also affecting their position in the protected home market.

Deregulation has advanced furthest in the *United States*, where the decision to split up AT&T, taken in 1982, became effective in early 1984.

Table 6.10   Participants in the deregulation debate on Telecommunications

Country	Category 1	Groups Category 2	Category 3
USA	AT&T Western Electric; seven Regional Bell Operating Companies	ITT; GTE/ SIEMENS Stromberg– Carlsson (PLESSEY); ERICSSON; MCI	IBM (PABX, satellites); Many others
JAPAN	Denden family NTT	AT&T (joint venture)	IBM/NIT; Matsushita and other Jap. elec. firms
FRANCE	CIT–Alcatel/ THOMSON; CGCT	ERICSSON; SIEMENS; AT&T– PHILIPS	IBM; MATRA
GERMANY	SIEMENS; SEL (ITT)	PHILIPS, ERICSSON	NIXDORF (PABX); IBM (Bildschirmtext, VANs); BOSCH, MANNESMANN;
UK	PLESSEY–GEC; STC, BT/ MITEL	THORN/ ERICSSON; AT&T– PHILIPS; Mercury	IBM (cooperation with BT) Matsushita (facsimile)

AT&T lost its monopoly in domestic telecommunications services, but became free to unfold activities abroad, while other companies got a chance to get a share of the large telecommunications equipment market in the USA, hitherto almost closed because of the AT&T preference for equipment of WESTERN ELECTRIC, its own captive producer.[92]

The *United Kingdom* followed the American example with the privatization of BRITISH TELECOM decided in 1982, and a licence given to a competing network (MERCURY) the same year. While BRITISH TELECOM has become privatized, it has not lost much of its monopoly position yet. The CABLE & WIRELESS subsidiary MERCURY Ltd is now competing for business customers, but 'its market share is unlikely to exceed 3% to 4% by 1990, and others cannot enter the arena until then'.[93] The legislative protection enjoyed by BT even extends to the particularly lucrative area of value added networks which allow the movement of information between far flung sets of incompatible equipment. Outsiders will not be able to use BT lines to offer such services until at least July 1989. Among other things, the breathing space will give BT time to sort out tariff systems that can stave off future competitors. 'The result is a kind of "private-sector monopoly" which . . . will end up about where AT&T was before the breakup . . . if privatization is all that happens to BRITISH TELECOM.'[94]

The *Italian* deregulation debate has to be understood against the background of Italy's fairly backward telephone network. The investment strength of the Italian telecommunications companies lagged due to the government's refusal to raise tariffs which caused nationalized companies such as the STET subsidiaries SIP and ITALTEL to operate at heavy losses in the second half of the 1970s. Tariff increases in 1981 and major financial injections from the government brought the companies in the black again but still left them without the financial muscle to make large-scale investments in new equipment. The government has initiated some form of privatization 'through the backdoor'[95] by lowering its share holding in SIP from 82.6 to 50.9 per cent. The entry of OLIVETTI (in cooperation with AT&T) and the efforts of IBM to team up with STET on Value Added Networks (VANs) have been welcomed as an attempt to facilitate the modernization of the Italian network. It may be a preview of the general IBM–AT&T battle in the years to come which will proliferate over the whole of Europe if governments do not find a suitable strategy for controlled deregulation in which the national firms can remain strong without becoming affiliates of either one of the two American multinationals.

[92] See Chapter Five on the activities of European producers in the USA; also Hills (1986).
[93] Piper (1985).
[94] According to Alvin Toffler, cited in Piper (1985), p. 40.
[95] See also Dang Nguyen (1985), p. 118.

In *France* the unquestioned monopoly of the Directorate General for Telecommunications (DGT)[96] is gradually coming under pressure. Although there is no question of privatization of the DGT, which is part of the Ministry of Posts and Telecommunications,[97] or allowing private companies to offer public telephone services, the Chirac government in May 1986 announced it would allow some competition in the area of Value Added Network services. This is the area in which firms like IBM and large users like the banks are trying to get a foothold in the tele-communications markets. Even before the change of government in France, therefore, IBM announced its intention to offer VANs in cooperation with the bank Paribas and the Semi-Metra consultancy. After the official announcement of a controlled deregulation in VANs, the groups around IBM on the one hand and around OLIVETTI (associated with the Banque de Suez) on the other were the first to apply for a licence. This immediately provoked speculations on France becoming another battleground for the IBM–AT&T (= OLIVETTI) rivalry.[98] Neither CGE nor BULL was (financially) able to enter this deregulated market as soon as the other firms did. Later BULL announced that it will cooperate with the American General Electric Corporation to ensure a share in this market.

The public monopoly has also come under heavy pressure because of the high rates for data transmission services to large companies, which belong to the highest in Europe, whereas for the smaller users they are among the lowest.[99] In 1986, however, the new centre–right government announced it would disentangle the cross-subsidization of local by long-distance phone calls, which will end the subsidization of private users by (large) business users.

A special case, again, is *West Germany* where pressures to limit the 'Fernmeldemonopol' of the Bundespost have remained limited, and the liberalization of the market for telecommunications equipment has not gone very far. A number of reasons can be advanced for this, among them the high degree of trade union organization of the Bundespost employees and the relatively advanced network of the German PTT. Another crucial factor is probably the constituency of the major ruling party in Germany, the Christian Democrats, in which representatives from small and medium sized industry (subcontractors of German telecommunications equipment producers) and from the region in which SIEMENS has its main

---

[96] Dang Nguyen (1985), p. 116.

[97] The DGT has an annual investment budget of FF30 billion, which is twice that of BRITISH TELECOM. Telecom Markets, *Financial Times Business Information*, 25 March 1986, p. 1.

[98] Eric le Boucher, Les télécommunications ouvertes à la concurrence, *Le Monde* (Paris), 17 May 1986.

[99] Commissie Steenbergen (1985), scheme 15.

production facilities (Munich) are very strong.[100] Also an important reason may be that major users of advanced telecommunications services (such as German multinational firms and banks), which elsewhere form an important pressure group to deregulate (category 4), have so many links with SIEMENS that there has been little domestic opposition to the company's privileged position as the major supplier.[101] Comparatively low rates for data transmission services to large customers may be an additional reason.[102]

While many countries thus take steps in the direction of deregulation, vested interests slow this movement down. A further liberalization of European telecommunications markets probably will have to be negotiated at the European Community level and will take considerable time. This would imply that the 'deregulation race' in this field will acquire a very peculiar feature: the participants will only run if they are assured that all will arrive at the same time!

The pace of deregulation in fields other than telecommunications will probably also be much slower than protagonists of far-reaching deregulation advocate. One of the reasons is that deregulation cannot go very far without creating grave risks. Any major hazards as a result of far-reaching deregulation could undermine the public acceptance of new technologies to such an extent that sales in the markets concerned could drop seriously. For this reason, multinational companies themselves are not a driving force behind general deregulation, but more behind general common standards to which firms from different countries would comply.

6.4. CONCLUSION

Deregulation must not be seen as a development towards a less close relationship between large corporations and national governments. On the contrary, to find a tailor-made mixture of regulation and deregulation that supports the interests of large companies best has become an important task for national political actors.

We can now try to give an answer to the questions raised at the beginning of this chapter. Have European high tech companies become less national as a result of the internationalization process, or have they become more dependent on government support? Can governments steer company

---

[100] Webber (1986).
[101] Lüthje (1985).
[102] Commissie Steenbergen (1985), scheme 15.

activities in core technologies, or do support programmes simply supply
to companies the finance they want?

In spite of the fast pace of the internationalization process, relations
between high tech multinationals and 'their' governments obviously are still
very close, with differences from sector to sector, though. Links between
government and multinationals are especially intense in the electronics sector,
but less important in a sector like the chemical industry. The multinational
companies of the electronics industry thus have a much more 'national'
character because they get the bulk of government R&D subsidies and are
strongly oriented towards the state market (defence, telecommunications),
whereas the chemical industry finances a larger share of its research itself
and sells only a small fraction of its production to governments.

The interdependence between states and high tech multinationals thus
remains strong, especially in the microelectronics cluster, but probably
less so in the biotechnology cluster which still has to develop. State and
large companies remain interdependent in so far as multinationals depend
on the state for the financing of research, orders, the basic education of
scientists and technicians, protection against foreign competition, and a
favourable regulatory framework, while the state depends on multinationals to improve the level of sophistication of 'national' production,
the competitiveness of the 'national' industry, the balance of payments
and the level of employment. This interdependence is lop-sided, however.
While companies may depend on public research subsidies, they often play
a significant role in formulating (and evaluating) the support programmes.
Where a large proportion of sales falls under public procurement, the
companies very often develop the standards that government departments
apply. A high share of public procurement in total sales therefore does
not necessarrily imply a one-sided dependence of the companies concerned
on the government. If the whole institutional and social context, in which
government–industry interaction is embedded, is taken into account, the
imbalance will often show to be the other way round, with companies
exerting more influence on the public administration than vice versa. This
can even be the case when the companies have been nationalized. The case
of France has shown that nationalization can be a means to channel even
larger amounts of public money to the companies than would otherwise be
possible, without a parallel increase in influence and control (see p. 197).
Where civil servants formulate programmes that do not take the
reservations of industrialists into account (like the plans to make France
conquer a 10 per cent world market share in biotechnology products, for
instance), the plans finally will be shelved (see p. 175).

Government–industry interaction in high technology concentrates very
much on large corporations. This remains the case even if explicit
programmes have been formulated to involve small and medium sized

enterprises more intensively. The political will to do so is not enough. The whole sociopolitical fabric would have to be changed in most countries in order to achieve this goal. Only a few countries have a socioeconomic structure (and a concomitant world market position) that allows for the inclusion of smaller companies in this relationship.

Inside Europe, government–industry relationships thus differ considerably from country to country. If we take into account the specific patterns of the relationships as analysed in this chapter, Germany, Sweden and Switzerland come out as very strong states with a far more balanced industrial and social tissue in which the government–business relationship is less predominantly confined to the large firms in the search for an 'appropriate' strategy in the international technology race (be it in subsidies, deregulation or otherwise). This corroborates the analysis in Chapter Five. The largest differences, however, do not exist between the larger European countries, in spite of all differences in the institutional settings and in the ideological debates on the role of the state in the restructuring process. They exist between the smaller countries and depend very much on whether 'national' (home-based) multinationals do exist in a country or not. It will be one of the topics of the next chapter to analyse whether national differences form a severe obstacle to closer cooperation among European high technology firms.

Chapter Seven

International Cooperation in Core Technologies

National efforts to support multinationals in core technologies in recent years increasingly have been supplemented by measures at the European Community level. The main argument for these measures is that the competitiveness of Europe's industry has to be defended against its rivals from Japan and America. European multinationals have played an important role in getting the European programmes off the ground. On the other hand, however, the very same multinationals cooperate intensively with their counterparts and rivals from these two countries in a myriad of cooperation agreements that have become so characteristic of high tech development. As a consequence, some of the results of European cooperation may 'leak out' to cooperation partners in Japan and the US—the very rivals whose success threatens the competitiveness of European companies. This chapter tries to explain this apparent paradox and to evaluate the implications of the pattern of cooperation that has come about.

Section 7.1 will describe how European technology policy was given shape. It will also describe the role which is assigned to the encouragement of cooperation among European companies. In section 7.2 the general reasons for international cooperation will be described in order to analyse to what extent cooperation among European companies and between European and American or Japanese companies are alternatives and can substitute for one another. In the third section, the pattern of cooperation among European companies in research and development is analysed in detail. This pattern is contrasted with the pattern of cooperation between European and American or Japanese companies in the fourth section. Do the different networks supplement each other? Is one the precondition for the other? Do European and global cooperation agreements contradict each other? Will global cooperation arrangements undermine the European technology policy? Will they help to boost the competitiveness of European firms, or will they condemn these firms to the role of junior partners of their American and Japanese competitors?

7.1. EUROPEANIZATION OF GOVERNMENT SUPPORT

7.1.1. Overlapping national efforts

European companies are no less research intensive than their American or Japanese counterparts (see Chapter Five). Neither do they lack size any longer. But, as described above, they are often more diversified than American companies and thus tend to spread their efforts over a larger area with the result that they often do not match the expenditures of their rivals in a given field. National governments have tried to compensate for these deficiencies by actively encouraging a restructuring and concentration of the industry and by supplementing company outlays for research and development by public efforts. There has been a certain shift in emphasis from traditional to 'new' ('sunrise') sectors, and most countries concentrate their support on more or less the same industries (see Chapter Six). It is not surprising, therefore, that the different national research programmes overlap to a considerable extent. While the national expenditures of European countries as a group often reach the level of American expenditures (and sometimes even surpass it), they are in practically all cases considerably higher than the Japanese efforts. This is the case if government expenditures are compared, or if government and industry expenditures are taken together. Obviously, however, the level of expenditures is only translated into competitive advantages in a few cases. A major reason is that large amounts of money are spent by different actors on the same kind of research, or that expenditures are spread so thin that a 'critical mass' for decisive breakthroughs is not reached. As a result, the effect of all the national efforts combined is much less than it could be if more coordination and cooperation did take place.

7.1.2. Counterproductive effects of national efforts

The national efforts are not only less effective than they might be, they can even be outright counterproductive. Parallel developments not only imply double (or even three-fold or more) investment, they also lead to different standards and other non-tariff barriers that make it difficult to realize a truly common market.

As has been shown in Chapter Six, a large share of total sales of the electronics industry goes to the state, and preferential public procurement is widely used to support national developments in the field of high tech. The close link between national producers and national governments, however, has a threefold negative effect on the development of new products. First, since most governments act in the same way, electronics companies in many lines of business are largely restricted to their national markets and thus are denied the economies of scale often necessary to earn the high investment in research and development during the short life-cycle of a new product. The second effect is that the guarantee of preferential treatment by national authorities decreases the incentive to develop a competitive design. Third, companies will cling longer to an outdated product in a small market, because they hesitate to switch to a new model before the investment in producing the first has been recovered.

Since the share of government expenditures in gross domestic product has increased considerably in recent years, and since preferences for national producers are even more common in the high tech sector than in other sectors, the Common Market increasingly gets undermined. Parallel national efforts thus not only constitute a considerable waste of resources, but from a European point of view they also impede the implementation of the Common Market. The restrictions on potential markets have a discouraging effect on companies that would have to conquer a very large market share in the restricted markets to make the development worthwhile.

7.1.3. The beginning of a European technology policy

Technology policy at a European level made a start with the foundation of EURATOM, the European Agency for Atomic Energy, as part of the Treaty of Rome which gave birth to the European Economic Community (EEC). Some contemporary observers at that time expected that EURATOM would be even more important than the EEC. Due to fundamental differences of interest, especially between France and West Germany, however, EURATOM never developed in the way originally intended. It is still highly relevant, however, for the actual technology policy of the European Community, because the vast expenditures on the nuclear energy research centres of the Community siphon away a large share of the means available for the support of research and development.

Until 1968, when a European *customs union* was realized, technology policy got little attention—not only at the European level, but at the national level of the member states as well. The European Federation of Employers (Union des Industries de la Communauté Européenne, UNICE), however, already in the mid 1960s was urging the European Economic Community to facilitate intercompany cooperation at a

European level and to relax the fiscal and juridical regulatory framework in order to stimulate industrial concentration and increase international competitiveness, first of all with regard to American companies.[1] After the European Community for Coal and Steel, EURATOM and the EEC had fused in 1967, the European Commission appointed a Directorate General for Industry (DG–III), and for Science, Research and Development (DG–XII), and prospects for closer European cooperation in key sectors were explored. An early result was the COST programme (Coopération européenne dans le domaine de la Recherche Scientifique et Technique) which started in 1971. This programme is somewhat similar to the later EUREKA programme in that it is not restricted to the member states of the EC (which do not have to participate if they are not interested in a specific project) and that its financing is shared by the countries involved according to their participation. Although some projects in the fields of informatics, telecommunications and new materials form part of the COST programme, most of the projects seem to belong to areas outside the field of core technologies, like traffic regulation, oceanography, meteorology, astronomy, agriculture, food technology, health and social sciences.

From the early 1970s on the Commission has been eager to expand its field of action and to start more ambitious technology programmes, while the larger member states have blocked any move into that direction. When the United Kingdom joined the Community in 1973, the member states' opposition to far-reaching European programmes was even strengthened. The 1970s was a period in which most European governments were active in drawing up their *national* R&D policies (see Chapter Six). This changed only in the early 1980s. At least three reasons have been responsible for this policy change:

(a) By 1980, the potential for national coordination and concentration seemed to have become exhausted. Where necessary and feasible, national companies had fused or concluded cooperation agreements, but international competitiveness often had not improved as expected. To mobilize additional resources and knowhow, cooperation had to be extended across national borders.

(b) The national technology programmes had proved insufficient to improve international competitiveness decisively. Their obvious overlap made it mandatory to try to come to a minimum of coordination if not to common programmes.

(c) The competitive position of European industry had worsened even more, especially in the crucial field of microelectronics.

---

[1] Swan (1983), p. 135.

7.1.4. The commitment of European multinationals to a European approach

The last mentioned factor explains why new initiatives in the field of information technology were undertaken first. Already in 1976/7, representatives of European information technology companies and national governments had come together and over a period of nine months discussed the future of European microelectronics.[2] A common long-range plan in information technology was suggested but did not get off the ground as the large member states insisted upon strict budgeting rules (commitments on a year-to-year basis only). This impeded the realization of long-term initiatives. Pressure from European electronics companies, however, increased. In the late 1970s, the twelve largest European electronics companies formed a 'Round Table' group with Étienne Davignon, then European Commissioner for Industry. These twelve companies are: GENERAL ELECTRIC COMPANY, ICL, PLESSEY, THOMSON–BRANDT, CIT–ALCATEL (CGE), BULL, SIEMENS, AEG, NIXDORF, OLIVETTI, STET (IRI) and PHILIPS. Though it proved difficult in the initial meetings for these twelve companies, which fiercely compete with each other, to agree, a kind of informal 'code of conduct' developed which made fruitful discussion possible, without putting the major interests of any of the participants at risk.[3]

The fact that some internal contradictions had to be overcome explains why it took several years before common action was taken. In 1983 the twelve sent a pressing letter to the Commission in which they presented a dramatic view of the competitiveness of the European information technology industry and underlined that even the national programmes of the large European countries were not sufficient to safeguard Europe's position in the long run.[4]

The European Council of Ministers thus came under pressure from two sides. The plans of the Commission were supported by a strong lobby from the large companies which could influence their national governments — with the result that the European Strategic Programme for Research and Development in Information Technology (ESPRIT) was accepted in record time by the Council.

But the large electronics companies not only exerted political pressure; they also contributed in a decisive way to shaping the ESPRIT programme.

---

[2] See *Financial Times*, 22 March 1977.
[3] Pannenborg (1986), p. 25.
[4] Cited in Commission of the European Communities (1983a), p. 3.

In a statement before the House of Lords, Mr D. H. Roberts, Technical Director of GENERAL ELECTRIC COMPANY, described this process:

> . . . first thing that happened was — that the Round Table established a Steering Committee; inevitably the Steering Committee established working parties because that is how we do things and these led to the technical panels. There were five of these where, in general, very good technical people from the 12 participants spent a lot of time together, formulating in outline the shape of the technical programme. I find it very difficult, as a member of GEC or any other of the 12 companies, to say that we do not think the shape of the programme as defined was sensible because we had excellent opportunity to influence it and in many areas I think we did. . . . it is not a programme dreamt up by Brussels bureaucrats and forced on us, it is our programme.[5]

About 100 employees of the twelve electronics companies cooperated in several commissions for a couple of months to work out the details of the ESPRIT proposals. Since the newly formed 'Information Technology Task Force' of the European Commission lacked detailed knowledge of the different research fields to be strengthened by the ESPRIT programme, the Commission had to fall back on the companies themselves to formulate the final programme. To monitor the execution of the programme, an ESPRIT Advisory Board (EAB) was formed, of which half of the members were representatives of the large electronics companies.

In the same year in which the twelve electronics companies wrote their urgent letter to the Commission, another group of European industrialists was formed which was not restricted to one industrial branch only — the *Round Table of European Industrialists*, also known as the Gyllenhammer group because it is presided over by VOLVO's chairman, Pehr Gyllenhammer.[6] It serves a double interest. On the one hand, it provides the large European companies with a forum in which they can discuss and

---

[5] House of Lords (1984), p. 37.

[6] Originally dubbed G–17 because its membership initially totalled 17 men, the Roundtable by July 1987 had 29 members: Pehr Gyllenhammer (VOLVO), Umberto Agnelli (FIAT), Wisse Dekker (PHILIPS), Torvild Aakvaag (NORSK HYDRO), Carlo de Benedetti (OLIVETTI), Jean-Louis Beffa (SAINT-GOBAIN), John Clark (PLESSEY), Étienne Davignon (SIBEKA SA, Belgium), Kari Kariamo (NOKIA, Finland), Karl-Heinz Kaske (SIEMENS), Herman Maucher (NESTLÉ), Hans Merkle (BOSCH), Curt Nicolin (ASEA), Anthony Pilkington (PILKINGTON BROS, Great Britain), Stephan Schmidheiny (ANOVA, Switzerland), Luis Solana (COMPANIA TELEFONICA NACIONAL DE ESPANA), Dieter Spethmann (THYSSEN), Poul J. Svanholm (DE FORENEDE BRYGGERIER, Denmark), Josef Taus (CONSTANTIA INDUSTRIEVERWALTUNGS GmbH, Austria), Pierre de Tillesse (PIRELLI), Werner Breitschwerdt (DAIMLER–BENZ), Raul Gardini (FERUZZI), Alain Gomez (THOMSON), Patrick Hayes (Waterford Glas), Luis Magana (Furnas Electricas de Cataluna), Jerome Monod (Lyonnaise des Eaux), Patrick Sheehy (BAT Industries).

formulate ideas for political action at the European level. On the other hand, it is a sounding board for the European Commission, and its members act as a strong lobby within their own national context in favour of European initiatives. While Gyllenhammer has been described as the driving force behind this Round Table group, it was in fact the EEC's former industry commissioner, Étienne Davignon, who recruited most of its members. 'Davignon had wanted a sounding board for his own ideas for reshaping European industry, and at the same time he wanted to establish an ongoing dialogue with people who could genuinely claim to be the voice of industry.'[7] But the Commission's chief perception of the Round Table is as a group of lobbyists who they hope will argue Brussel's case on various industrial issues at the highest level in each European capital.[8]

The Round Table has published three reports so far. The first, in December 1984, was called *Missing Links* and argued the need to allocate $60 billion over 20 years for large traffic infrastructure projects. The second report, *Changing Scales*, published in June 1985, was a survey of the economies of scale that would result from a genuine unification of the Common Market. The third report, *Making Europe Work*, deals with proposals for tackling Europe's deepening unemployment crisis. The Round Table's most concrete activity has been the creation of the Amsterdam-based venture capital company, EUROVENTURES BV, with an initial capitalization of $30 million, that has been subscribed by ten of the member companies. Interestingly enough, some former members of the group have left the Round Table: Peter Baxendeel of ROYAL DUTCH SHELL, Kenneth Durham of UNILEVER, John Harvey-Jones of ICI and Antoine Riboud who heads BSN–GERVAIS DANONE. While Harvey-Jones of ICI is said to have left because the Round Table in his opinion had failed to focus on Europe's fundamental issues (the need, for instance, to overhaul the restrictive EEC competition laws, and the case for accelerating the EEC timetable for creating a genuinely common market), the chief executive officers of SHELL and UNILEVER may have left because their companies' interests are less specifically Europe oriented than global.[9]

Alongside the Round Table of European Industrialists (G–27), the Round Table of the twelve European electronics companies that have given birth to the ESPRIT programme continues to exist. Eight of the twelve companies also formed the *Standard Promotion and Application Group*

---

[7] Merritt (1986), p. 23.

[8] According to Bruno Liebhaberg, adviser for industrial and technology questions in the Cabinet of Delors in Brussels, 'these men are very powerful and dynamic . . . when necessary they can ring up their own prime ministers and make their case'. Merritt (1986), p. 22.

[9] Merritt (1986), p. 26.

Table 7.1   Companies most involved in European Affairs

Firms	G–27	EUROVENTURES	ESPRIT	SPAG
PHILIPS	×	×	×	×
SIEMENS	×	—	×	×
NIXDORF	—	—	×	×
BULL	—	—	×	×
THOMSON	—	—	×	×
ICL	—	—	×	×
STET	—	—	×	×
OLIVETTI	×	×	×	×
GEC	—	—	×	—
PLESSEY	×	—	×	—
CGE	—	—	×	—
DAIMLER/AEG	—	—	×	—
VOLVO	×	×	—	—
FIAT	×	×	—	—
SAINT-GOBAIN	×	×	—	—
RENAULT	×	—	—	—
MATRA	×	—	—	—
NESTLÉ	×	—	—	—
BOSCH	×	×	—	—
ASEA	×	×	—	—
CIBA–GEIGY	×	—	—	—
PIRELLI	×	×	—	—
Others	13	2	—	—

(SPAG) in 1985, with the aim of unifying European standards and with one voice representing European interests in international standardization organizations. Table 7.1 indicates the membership of European companies in these different groups.

Two companies participate in all the activities summarized in Table 7.1, namely PHILIPS and OLIVETTI. Both companies have strongly backed any step towards a more integrated Europe. At the same time, both companies have concluded far-reaching agreements with (among others) the *American* company AT&T and have therefore been accused of betrayal of the European cause. The discussion on whether European cooperation agreements are alternatives to agreements with American and Japanese companies will show whether this perception is justified or not.

Before reviewing the networks of European cooperation agreements and their origin, and comparing the network of agreements concluded by European companies with American and Japanese firms, we want to describe the most obvious reasons that lead to cooperation agreements, especially in the field of advanced technologies.

7.2. REASONS FOR INTERNATIONAL COOPERATION

International cooperation among companies — whether in the form of joint ventures, licensing, minority shareholding, cooperation in research, production or marketing, contract research, etc. — has existed for a long time. However, the economic crisis of the 1970s and the early 1980s led to a new wave of cooperation agreements. These agreements have by no means been restricted to high technology activities, but there is no doubt that the network of cooperation arrangements that has been created in this field is especially intense. There are at least six reasons behind this recent phenomenon.

7.2.1. Risk sharing

A very important reason is risk sharing. New developments in the field of high technology entail a number of risks. It is often also not certain whether a specific technology will be feasible, or whether it can be produced in a profitable way. Second, there is always the *risk that a competitor is faster* and will get his result patented, which would make it difficult for other companies to exploit the results of their parallel research. Third, there is the *risk of patent litigation* that can spoil the market conditions even for the company that succeeded first in a specific development. AKZO and DUPONT are examples in this context. Each had developed a new synthetic fibre at very high cost (see Chapter Two), but found the market halved when the other company accused the firm of having infringed its patent rights. Fourth, pending patent litigation can increase the *risk of market access.* This also includes the danger that the product is not licensed by the regulatory authority in question. Such a license is, for instance, required for many biotechnology products. Fifth, there is the risk of not getting a market share large enough to cover R&D investment and production costs. Sixth, there is the problem of *market acceptance.* It is not only necessary that a new product is produced at a price low enough; the product has to be accepted by the public. This risk can be considerable as the history of single-cell protein for human consumption and the public discussion on the use of animal growth hormones show.

New discoveries often suddenly open up many new avenues of research simultaneously (the discovery of rDNA, for instance). In order to develop a specific kind of product as fast as possible, many different research paths may have to be tried. Any single company may not have the research potential to follow all these paths at the same time, and may therefore wish to cooperate with other companies.

High technology research is not only highly risky, it is also very costly in many cases. If markets are not large enough to recover high investment costs, the latter have to be kept within certain limits. This can be done, for example, by sharing the costs with other firms.

7.2.2. Market penetration in spite of protectionism

Protectionism is once again on the rise, especially for high technology products. Preferential public procurement is increasing as described above, and non-tariff barriers proliferate. Cooperation with a company inside foreign markets can circumvent many of these barriers.

7.2.3. Many-fold applications of core technologies

It has been stressed throughout this book that new core technologies are applicable in many different sectors and branches. This fact provides two additional reasons for cooperation agreements. One is that a new technology developed by a company probably will be applicable in several fields outside the company's traditional competence. The company therefore will have to look for partners if it wants to enter these fields, or if it wants to participate in the exploitation of its finding by other companies. Also, it is often the case that a company will gradually become aware that a firm from a different sector has developed a product or process that can be applied in the company's traditional fields of activity.

7.2.4. Complementary developments in different technology clusters

A closely related reason is that developments of a specific technology often can only proceed if an additional technology (or even several other technologies) is mastered. The increasing use of optoelectronics in telecommunications, for example, made it imperative for all major telecommunications producers to acquire technologies and knowhow in the field of optical fibres. The leading company in this field (possessing the most strategic patents), the American firm, Corning Glass, therefore now finds itself at the centre of an international network of cooperation agreements. The development of new products often requires different technologies from the same basic cluster as well as technologies from other clusters. The expected integration of biotechnology and microelectronics in 'bio-informatics' or 'biotronics', therefore, will probably lead to a host of new cooperation agreements between biotechnology and electronics companies that feel the need to complement their own background with experience in the other field.

7.2.5. The need for systems instead of stand-alone products

A central characteristic of the present wave of technological innovation is less the application of separate inventions than the *integration of a variety of different new products and processes into new systems*. Terms such as the 'Factory-of-the-future' and the 'Office-of-the-future' imply that further increases in productivity no longer depend on the introduction of one or the other specific machine such as a numerically controlled lathe, a robot, a word processor or a telecopy machine, but on the integration of a family of interrelated new machines which function in a complementary way to each other. With the introduction of 'intra-activity' automation (see p. 18), these 'families' will become even larger, comprising not only office technology, but also design equipment and production machinery.

A company can acquire considerable advantages if it can offer from the onset an entire family of products instead of only single items which remain to be integrated into complete systems. One advantage is that customers will not have to shop around but can acquire all they want in *one shop* with a single maintenance contract for all the equipment. Second, with the increasing complexity of systems, integrating individual machines into a single system has become an increasingly complex affair. Compatibility often cannot be fully assured. This makes equipment attractive that has been designed on the drawing board to be part of an integrated system.

Profitable as it is to offer comprehensive systems, few companies are able to cover the whole range of equipment required. Even if they do, there will always be new items of equipment produced by other companies which could make the system as a whole even more attractive for customers. Thus there is a continuous stimulus to conclude cooperation agreements with other companies, either to produce common equipment packages or to use another company's technology to produce equipment similar to that of another firm for integration into one's own family of products.

7.2.6. Establishment of common standards

Another driving force behind international cooperation is the competition among firms to establish a common standard in the early phase of a product's development. This is not so much a form of risk sharing (see 7.2.1.) as of *risk aversion*. If different companies develop a similar product on the basis of different standards (as happened in the case of video-recorders), one standard eventually will emerge as the dominant one—with the consequence that most of the investment made in the development and production of competing models based on different standards will become obsolete.

Cooperation in the development of a common standard can also be desirable in order to achieve full acceptance of a new product by the public. As long as different standards compete with each other, potential buyers may hesitate to make a decision and may prefer to wait until a dominant standard has emerged. Maintenance services would have to be developed for the different types of equipment. As a result, any single type of equipment may not achieve the quality standard or have available to its users the density of service points as would have been achieved given the existence of a dominant standard.

A reason for cooperation in establishing common standards can also be that a number of smaller companies may try to formulate alternative common standards in order to try to exclude a dominant competitor from certain markets. An example here is the European computers producers in their efforts to create a common standard in order to fight off IBM.

7.2.7. The dependence of competitiveness on the right cooperation agreements

Cooperation has become a strategic tool to improve a company's competitive position. The decision on the right cooperation partner can be more important than the decision on which technology to pursue. Some European companies have handled this instrument with some virtuosity. A case in point is OLIVETTI which does not have a reputation as a technological high-flyer and does not spend as much on research and development as competing companies. Under the chairmanship of De Benedetti, however, the company embarked upon a comprehensive strategy to enter into a great number of alliances with well-established companies. This strategy has proved to be financially highly beneficial.

The success of this alliance approach has encouraged other companies to follow a similar strategy. This is partly due to the 'power of example' and also to the changed pattern of international competition which are forcing companies to follow suit. Once its competitors enter into a strategic alliance, the position of a company may suddenly change—indeed, more rapidly than in the case of changes brought about by competing technological developments. In order to avoid unhappy surprises and to secure the most attractive cooperation partners before they become engaged in agreements with competitors, some companies have adopted a policy of the 'preemptive strike', that is signing a cooperation agreement expressly in order to deny a competitor an attractive partner.

European companies, however, used to be somewhat less experienced and less eager to conclude international cooperation agreements than American and Japanese companies, partly because of the more diversified structure of European firms. Being more diversified, their chances were greater of finding an 'in-house' solution to a given problem by turning to another division instead of to an outside company. European companies also lacked the experience Japanese companies accumulated as a result of their many licence agreements with American firms to import US technology. Europe acquired American technology usually through American foreign investment in Europe. Also, European firms have not been under pressure to conclude cooperation agreements in order to penetrate foreign markets, as have Japanese companies which have had to deal with increasing barriers erected in response to the fast growth of Japanese market penetration. The smoother growth of European exports has protected Europe, relatively speaking, from the anti-Japanese type of response.

These differences in the inclination to conclude cooperation agreements are only differences in degree, not in kind. The threshold for European companies to enter a cooperation agreement may just be somewhat higher than for American or Japanese companies. This may be one of the reasons why European companies reach more agreements with American or Japanese partners than with other European firms. Negotiations with an American or Japanese prospective partner may be more straightforward and less filled with mutual mistrust than is the case with two European prospective partners. Negotiations between European firms can also be more difficult because their markets may overlap to a larger extent. As a result competition may dominate in their relationship leaving little room for cooperation. The pattern of cooperation agreements with other European companies and with American and Japanese companies respectively will be analysed in more detail in sections 7.3 and 7.4.

7.3. EUROPEAN NETWORKS OF COOPERATION

In this section the pattern of cooperation among European companies will be examined with specific reference to a number of the industrial cooperation programmes initiated by the European Commission. The background of the four most relevant programmes is described in 7.3.1. In most of the programmes considered the large companies play a dominant role (7.3.2). Finally, the patterns of cooperation emanating from these programmes will be considered (7.3.3).

7.3.1. The European programmes

In many respects the example of temporary multilateral pre-competitive cooperation on R&D provided by the Japanese VLSI project shaped the model for European cooperation programmes. The first of these, the ESPRIT programme, aimed also at pre-competitive research cooperation in which the participating firms and the European Commission each pay 50 per cent of the project costs. A special Information Technology Task Force was created within the European Commission (next to DG–III) to steer the programme.

The political success of ESPRIT provoked a number of reactions. Another Directorate of the European Commission, DG–XII (Science and Technology), decided to launch BRITE (Basic Research in Industrial Technologies in Europe), its own pre-competitive research programme aimed at industrial applications in less advanced sectors, such as textiles. To organize support the directorate called for endorsement by a larger group of companies than the twelve of ESPRIT. Hundreds of large and smaller companies responded positively.[10] This strengthened DG–XII's bargaining position *vis-à-vis* the Commission and the Council to such an extent that a pilot phase could get started by 1985. BRITE is mainly aimed at R&D on new materials and production automation and is more oriented towards commercialization than is ESPRIT. This makes it also more suited for participation by smaller companies. Additional projects in new materials and biotechnology were launched by the European Commission. The less than encouraging results of Commissioner Davignon's efforts to create an ESPRIT-like Round Table for biotechnology have already been discussed in Chapter Six. The Information Technology Task Force, which by then had already become more bureaucratized and weighed down with

[10] Interviews.

a large number of political appointees, announced a programme to develop a European broadband telecommunications network in response to similar Japanese and American projects. Dubbed RACE (Research and development in Advanced Communications technology in Europe), the programme began to take shape with the 1985/6 pilot projects. The ESPRIT 'founding fathers' again played a prominent role. But the laboratories of the national PTTs also actively participated. RACE thus originated from the same source as ESPRIT, while BRITE developed from a different source. Placing assorted European cooperation projects under one name, therefore, obscures differences of interest within the European Community. *A fortiori* that is the case of the last large European technology project considered in this chapter: EUREKA (European Research Coordination Agency). EUREKA was presented by the French as an alternative to the Strategic Defence Initiative ('Star Wars') of US President Reagan. Such a plan had already been proposed and shelved in France a number of years before, but was again suddenly dusted off when US Secretary of Defense Weinberger gave the European countries and Japan 60 days to formally respond concerning whether or not they wanted to participate in SDI. The French Government sought to prevent the participation of German and British industry, arguing that it would further weaken Europe's technological position. EUREKA is officially aimed at creating a civilian alternative to SDI, consisting of directly marketable products from companies that might otherwise participate in SDI. EUREKA also became an alternative to the involvement of the European Commission and a framework deemed suitable for pre-competitive large cooperation projects. In this sense EUREKA affects the authority of the European Commission in competition policy issues and is thus contributing to the deregulation race.[11] EUREKA also leads the European Commission to press for a 'European Technology Community' and to seek more funding and authority *vis-à-vis* the European Council of Ministers.

As discussed in the section in Chapter Six on national subsidy programmes, an important characteristic of the European technology projects is their *reactive* nature: interest groups react to events in the USA and Japan, and to initiatives taken in other parts of the Community. The interest groups then search for coalition partners in order to reinforce their own bargaining position. The fact that EUREKA so closely resembles COST (see 7.1.3) received little attention in the internal debate. Creating something 'new' with dispatch took priority.

Although the R&D subsidy of the Community is quite small compared to the total R&D budget of the large companies (the total funding of

---

[11] Schutte *et al.* (1986), p. 20.

ESPRIT, RACE and BRITE together is less than, for example, the annual research budget of SIEMENS), the financial incentive for companies is none the less an important reason to participate. In the first place, the projects provide cheap 'windows on technology' for the participating firms. With around 30 projects in ESPRIT, PHILIPS, for example, acquired 166 'windows' on industrial and scientific research in Europe. With the input of around 600 man–years (worth more than DFL100 million) the firm taps a reservoir four times as large, namely 2300 man–years.[12] Second, matching subsidies covering 50 per cent of the costs of each project represent a considerable incentive and can make a difference in a firm's decision to undertake less attractive, 'second-best' projects. Third, some of the international cooperation agreements would have been started anyway. In such cases additional financial support is welcome, provided that it does not mean the involvement of other competitors and/or cumbersome bureaucratic control. The European Commission has tried to minimize bureaucracy. If we consider the ESPRIT projects on chip technology in which PHILIPS and SIEMENS are participating, they are often the sole industrial participants in these projects. But their more important (Megachip) project is being organized outside the framework of the European Community programmes. Clearly both firms want to prevent a collaborating firm from acquiring their knowhow in technologies the importance of which goes far beyond the range of specific ESPRIT projects.

The precise extent to which there is overlap between the various European industrial cooperation programmes is unknown, although specific projects do resemble each other very much. An attempt is being made to coordinate different efforts in order to prevent duplication. The duplication of national and European programmes remains considerable, however, because all European governments have national projects comparable to those of the European Community (see Chapter Six). According to the directorate of the Alvey programme in the UK, for example, the overlap between Alvey and ESPRIT is 25–30 per cent.[13] Coordination becomes especially difficult in this area as a result of the ongoing debate between the European Commission and the smaller as well as the larger countries on the question of whether European projects should be substitutes for national programmes or merely complementary. The governments of the larger countries favour the latter. The growing battle between the different groups within the European Community has already seriously frustrated the decision-making process on the new technology framework programme for the late 1980s and early 1990s.

---

[12] Dr P. Kramer (Philips), presentation, Efficiency Beurs, Amsterdam, October 1986.
[13] House of Lords (1984), p. 36.

7.3.2. The dominance of large companies in European programmes

As has been noted earlier, 'the twelve' have played an important part in the preparations for the ESPRIT programme. They are also actively engaged in monitoring the progress of the programme. From the Round Table to the panels responsible for the working plans on the five subprogrammes of ESPRIT (Advanced Microelectronics, Software Technology, Advanced Information Processing, Office Systems and Computer Integrated Manufacturing), managers and technicians from the twelve are heavily involved. The formal channels of influence of European industry in ESPRIT are almost completely dominated by the twelve large electronics companies. Next to these institutions, an Advisory Board, the Council of Ministers, and the Management Committee consisting of representatives of the national governments, participate in the decision-making process around ESPRIT. On the basis of a qualified majority, the Management Committee evaluates projects exceeding ECU 5 million. This has been one of the few areas where member states have departed from the *de facto* unanimity principle usually applied in European decision making. It accelerates the decision-making process, but can work to the disadvantage of the smaller countries. If the large countries are in favour of a project, it will be approved over any objections of the smaller countries.

Although the selection process in ESPRIT is supposed to be carried out on the basis of the merit of the projects submitted, it is hardly surprising that the twelve are very well represented in the approved projects. Since they have helped to define the framework of the programme, the projects are often tailor-made for them. In 1985 in 70 per cent of all ESPRIT projects one or more of the 'founding fathers' participated. Table 7.2 gives some figures on relevant groups of participants in the four major European cooperation projects.

Included are the projects up to mid 1986. ESPRIT-a and ESPRIT-b stand for the first and second phase in this programme. The pilot phase of ESPRIT has been left out of the table, whereas only the first phase of the BRITE programme and the pilot phase of RACE could be considered. With regard to EUREKA, most of the projects approved at the London conference in June 1986 have been included.

With regard to ESPRIT, it is noteworthy that the twelve did not participate enthusiastically in the CIM subprogramme which was added by the Commission. Consequently, this part of the ESPRIT programme has lower funding—13 per cent of the total budget, whereas other subprogrammes account for more than 20 per cent of the budget—and the CIM secretariat had great difficulty in even keeping this budget from

Table 7.2   The dominance of large companies in European projects

	EUREKA*	ESPRIT-a	ESPRIT-b	BRITE	RACE
A. Presence of 'the twelve'	22%	74%	69%	23%	66%
B. Presence of one of the sample firms	47%	79%	87%	49%	91%
C. Presence of other multinationals	23%	36%	26%	23%	53%
D. Presence of US and Japanese multinationals	4%	15% 11%‡	10% 6%‡	6% 4%‡	44% 22%‡
E. Exclusive presence of SMEs and research organizations	13%†	8%	6%	33%	6%

*One third of EUREKA projects are still open to new participants. In around 10% of the projects, the participants are unknown.

† At the London conference in June 1986, the degree of exclusive participation of SMEs and research institutes proposed was more than 30%.

‡ Since the acquisition of ITT by CGE, the participating former ITT subsidiaries have been considered to be 'European' and the presence of US multinationals consequently has diminished considerably.

being squeezed. This may be due first of all to the fact that the twelve did not initiate this part of the programme. Second, one of the CIM project's goals is to *integrate* technologies. This will more directly affect the short-term competitive position of the European multinationals. Besides, many non-electronic European firms such as automotive and machine tools firms (see Chapter Two) are strong competitors in this area. Third, IBM had targeted this part of the ESPRIT project and had submitted most of its proposals for European cooperation under this heading. Since one of the main reasons behind ESPRIT was to find an effective answer to the threat of IBM dominance, it is hardly surprising that firms were reluctant to participate. Whether this has been deliberately organized this way or is the result of uncoordinated action remains to be seen. But it is noteworthy that in two out of the three projects in which IBM participates, the projects have the *largest* number of participants (sixteen and nineteen respectively) of any ESPRIT project. In EUREKA, IBM does not cooperate with any of the EC electronics firms, but with two Swiss firms. IBM, on its side, is only willing to cooperate in areas complementary to its traditional core activities in mainframe computers (CIM in ESPRIT, expert systems in EUREKA, communications software in RACE). IBM clearly does not want to transfer technology in areas where it is dominant. But European companies tend to behave similarly to IBM in this respect. US multinationals are not excluded from the cooperation programmes. But, with the exception of ITT telecommunications activities and AT&T, whose joint venture with PHILIPS had raised the presence of US multinationals to more than 40 per cent of all RACE projects in 1986, US participation remains limited. After the 'Europeanization' of the control over the telecommunications department of ITT, the share of US presence in RACE projects became half. For firms such as DIGITAL EQUIPMENT, being included in the European cooperation projects contributes to their image as true European corporations. The almost total absence of Japanese multinationals is apparently due to their low 'European' profile.

In ESPRIT, the participation of small and medium sized companies (SMEs) has been labelled 'representative' for their involvement in research and development activities.[14] Since the bulk of R&D in the industrialized countries is done by the large firms, the financial participation of SMEs is marginal, despite their participation in half of ESPRIT projects. Only in 6 to 8 per cent of the cases do SMEs participate on an exclusive basis with each other or with research organizations. In all the other cases, they cooperate with large firms as 'junior partners', running the risk that their senior partner will make greater use of their knowhow than vice versa. ESPRIT strengthens the status quo in Europe with a few large multinationals doing the bulk of R&D in core technologies. The argument concerning the difficulties of marshalling effective support for smaller companies, as discussed in the previous chapter, can thus also be applied to the European level. A typical rationalization of this phenomenon points to the circumstance that only the large European firms need support for their competitive position, especially towards the Japanese danger: 'we are not being damaged by *small* Japanese companies'.[15]

Proposals of the European Parliament to raise the subsidy share for companies with less than 500 employees to 65 per cent and to lower the subsidy share for the large companies to 35 per cent of project costs, in order to expand participation of SMEs, have not been effective. The European Commission's coolness to such proposals is due to their fear that they might alienate the large IT companies in Europe from the programme and thereby weaken the Commission's bargaining position towards the member states in general and towards the European Council in particular.

The participation of the twelve in the RACE programme is as large as their participation in ESPRIT. In EUREKA, participation of the twelve

---

[14] Pannenborg (1986), p. 27.
[15] Mr Roberts (GEC), statement before the House of Lords (1984), p. 50.

Table 7.3   Number of European projects in which European multinationals participate (date: June 1986)

Firm	EUREKA	ESPRIT-a	ESPRIT-b	BRITE	RACE	Total
SHELL	0	0	0	0	0	0
BP	0	2	2	1	0	5
IRI/STET	1	15	15	3	7	41
UNILEVER	0	0	0	2	0	2
ELF	1	0	0	2	0	3
PHILIPS	5	17	14	4	9	49
SIEMENS	2	15	10	8	3	38
VOLKSWAGEN	0	0	1	1	0	2
DAIMLER/AEG	3	18	12	4	4	41
BAYER	1	0	1	0	0	2
HOECHST	0	1	2	0	0	3
RENAULT	1	3	0	6	0	10
FIAT	2	7	3	4	3	19
NESTLÉ	0	0	0	0	0	0
BASF	1	1	1	1	0	4
VOLVO	1	0	0	1	0	2
ICI	1	1	0	3	0	5
PEUGEOT	2	0	1	4	0	7
CGE	5	9	12	4	4	34
SAINT-GOBAIN	1	0	0	0	0	1
CIBA–GEIGY	1	1	0	2	0	4
DSM	1	0	0	2	0	3
GEC	3	18	21	7	12	61
KRUPP	0	1	1	3	0	5
THOMSON	6	13	20	0	8	47
GR. MET.	0	0	0	0	0	0
RHÔNE–POULENC	3	0	1	1	0	5
BOSCH	1	0	2	1	3	7
MANNESMANN	0	0	0	0	2	2
AKZO	1	0	0	2	0	3
DALGETY	0	0	0	0	0	0
THORN–EMI	1	1	1	0	3	6
ASEA	1	0	0	0	0	1
ERICSSON	0	0	1	0	2	3
OLIVETTI	2	13	14	1	1	31
BULL	3	19	14	0	0	36
ICL/STC	1	10	17	0	5	33
DEGUSSA	0	1	0	2	0	3
GIST BROC.	0	0	0	0	0	0
BRITISH TEL.	0	6	5	0	11	22
MATRA	4	4	4	0	3	15
NIXDORF	1	8	3	2	1	26
PLESSEY	1	13	4	0	4	22
IBM	1	2	1	0	2	6

(continued)

Table 7.3   (continued)

Firm	EUREKA	ESPRIT-a	ESPRIT-b	BRITE	RACE	Total
ITT	0	15	4	2	14	35
DIGITAL EQ.	0	1	2	0	0	3
GTE	0	0	0	0	4	4
ROCKWELL	0	0	0	2	0	2
UTC	0	0	1	0	0	1
HP	0	0	0	0	1	1
HONEYWELL	1	2	0	1	0	4
FOXBORO	1	0	1	0	0	2
ASAHI GLASS (JAP.)	0	0	1	0	0	1
FORD	0	0	0	1	0	1
APT	0	2	1	0	4	7
Total projects	72	104	94	95	32	397

Table 7.4   Origin of companies participating in European projects

Country	ESPRIT-a	ESPRIT-b	RACE	BRITE	EUREKA	Average
France	64%	71%	75%	46%	67%	64.6%
UK	66%	68%	81%	53%	37%	61.0%
Germany	65%	62%	50%	58%	29%	52.8%
Italy	47%	45%	53%	25%	22%	38.4%
Netherlands	27%	30%	44%	26%	16%	28.6%
Belgium	20%	23%	25%	20%	15%	20.6%
Denmark	14%	17%	22%	14%	8%	15.0%
Spain	0%	6%	16%	0%	21%	8.6%
Ireland	14%	10%	9%	8%	0%	8.2%
Greece	4%	6%	9%	8%	1%	5.6%
Sweden	0%	0%	3%	0%	11%	2.8%
Switzerland	—	—	—	—	14%	2.8%
Portugal	0%	0%	3%	0%	6%	1.8%
Finland	—	—	—	—	8%	1.6%
Austria	—	—	—	—	8%	1.6%
Norway	—	—	—	—	5%	1.0%
Luxemburg	0%	0%	0%	1%	1%	0.4%
Turkey	—	—	—	—	1%	0.2%

is much lower. Nevertheless, the general dominance of multinational companies resembles that of RACE and ESPRIT. The BRITE programme, however, provides a considerably different picture. The involvement of the twelve is relatively small, whereas the share of SMEs and research institutions participating *without* any presence of multinational corporations is much higher than is the case on the average in the other projects. About a third of all BRITE projects belong to this category.

As in the case of the CIM part of ESPRIT, BRITE's problem definition and the general framework of the programme have not been formulated by the twelve. This has meant less initial influence on the programme and also a lesser degree of participation. The BRITE programme is also the only project in which some, albeit indirect, involvement of European trade unions has been achieved.

The participation of the European multinationals from our sample in the four European cooperation projects is listed in Table 7.3. By mid 1986, the champion in European cooperation, present in one out of every six projects, is the British GENERAL ELECTRIC COMPANY. Paradoxically, GEC, which is one of ESPRIT's 'founding fathers', is not known as a firm with a strong European commitment. GEC, whose direct interests still seem to be predominantly in the UK, is not participating in such initiatives as the G-27 group, EUROVENTURES or the SPAG group. Neither is GEC inclined to support greater liberalization in Europe since this could negatively affect its already weak position in the British market. Indeed, GEC's high degree of participation in European R&D projects can be viewed in this context as a *substitute* for cooperation on more far-reaching projects on a European scale.

Other firms scoring high on participation in the European cooperation programmes all come from the electronics sector and have had a central position in the clusters of core technologies which have been discussed in Chapter Two. PHILIPS, IRI-STET, DAIMLER-AEG and THOMSON make up the first layer, participating in more than 40 projects, and SIEMENS, CGE, OLIVETTI, BULL, ICL/STC, and, surprisingly, ITT the second layer, with 30-40 participations. Most of the ITT participations are related to its telecommunications activities, which after its 1986 takeover by CGE fell to the latter's total. CGE has thus become as highly involved in European cooperation programmes as GEC. Interestingly, the French firm is, like GEC, not an active participant in other institutions pressing for European unification. This somewhat paradoxical fact may merit further research. Most other first and second layer firms are actively promoting a politically and economically more united European Community.

Petrochemical, pharmaceutical and food-processing firms have the lowest number of participations. Furthermore, they have not been eager to support any strong European cooperation strategy in other areas (except in the exceptional case of a crisis cartel in chemical fibres). The large automobile producers have been fairly active, with the notable exception of VOLKSWAGEN. From the group of electronics firms, BOSCH and THORN-EMI have been reluctant to cooperate with other European firms, thus remaining true to their general philosophy of 'going it alone'.

7.3.3. Patterns of cooperation

The EC projects have helped to intensify cooperation in research and development between European firms. In considering the country of origin of the participants (Table 7.4) the dominance of French companies in EUREKA by the end of 1986 stands out. In two thirds of the projects French firms participate. This is due not only to the fact that the French launched the project, but also that the French Government was the first to promise large-scale funding for French participants.

In the EUREKA programme, the firms and institutions from the smaller European countries have a more marginal position than in the other programmes. In the programmes which are coordinated by the European Commission the principle of 'just retour' (referring to giving countries a share in the benefits of programmes exactly as large as their relative contribution to the whole EC budget) has had some effect on the degree to which the smaller countries can participate. The Dutch contribution might be considered to be more than a 'just retour'. This relatively good position of the Netherlands must be attributed to the strong involvement of PHILIPS. If we leave out the PHILIPS–Holland participations, the relatively high share of Dutch participation drops below that of Belgium and comes closer to that of Denmark.

The BRITE programme has a distinct position due to its more equitable distribution of projects among participating countries. Table 7.4 also puts into relief the markedly weak positions of Greece and Portugal. This is clearly due to their lack of core firms. This corroborates the analysis of the regional impact of new technologies made in Chapter Four. Significant in the case of Greece is that the University of Athens has been almost the sole actor in cooperation agreements.

As a result of the European programmes a network of cooperation relationships in research and development has developed which did not exist before 1983. The densest segment of this network centres around the IRI/STET group and GEC, which jointly participate in at least 30 projects. In many of the projects in which the other core firms in the network (such as SIEMENS, PHILIPS, THOMSON, DAIMLER/AEG and CGE) participate IRI or GEC is also involved. Europe's most diversified electronics multinationals are thus closely linked to each other.

One requirement of the European programmes has been that firms of different nationality participate. As a result the network links international actors more than national actors. Many of the participating firms, however, have used the European programmes to make additional cooperation arrangements with other national firms, especially in those projects with a large number of participants. GENERAL ELECTRIC,

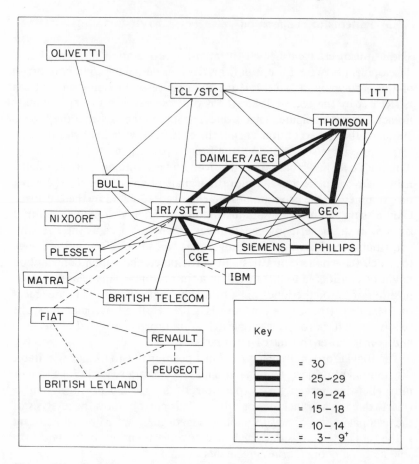

Figure 7.1   Europe's most intensely cooperating firms* in European R&D projects (1986)

*i.e. Three or more linkages in projects: Esprit (first and second tranch), RACE, BRITE, EUREKA.
†Excluding the linkages of the 'core firms' which, due to a few major projects in which they all participate, have more than 3 linkages with almost every other firm in the cluster. This would make the figure too cluttered. (The core firms are: IRI, Philips, Daimler, CGE, GEC, Thomson, Olivetti, Bull, ICL/STC, British Telecom, Plessey).

for example, has organized nine collaboration agreements with ICL and ten with PLESSEY.

An additional effect of the close network of European cooperation, more important than the inclusion of other national firms, however, might be the *exclusion* of other national firms from the network. The almost total lack of cooperation between IRI and OLIVETTI, or CGE,

THOMSON and BULL with MATRA (CGE has more links with IBM–France than with MATRA!) is a reflection of the large amount of national compartmentalization, especially in France and Italy.

It has been suggested that the network of cooperating large firms might pull the smaller national firms on the coat-tails of the large national concerns into the sphere of European cooperation and thereby counteract the exclusion of small and medium sized firms from European cooperation. The smaller firms were thus supposed to benefit from contacts with the larger firms. Up to mid 1987 this has not been the case. If we consider all the projects in which 'the twelve' cooperate, there are very few cases in which the large core firms 'pull' small national companies over the threshold of European cooperation. In the case of BULL, PHILIPS and THOMSON a *maximum* estimate of small firm participation from the same country comes to one out of three projects. In the cases of GEC, ICI, PLESSEY, NIXDORF, AEG, SIEMENS, IRI/STET, it is less than one out of ten projects. The large firms thus cooperate with national research institutes and universities, and cooperate in many instances with other large national firms, but for the most part deny smaller firms the opportunity to benefit from their cooperation. This comes close to excluding the smaller firms from the more important collaboration agreements and has created a pattern characteristic of the large European cooperation programmes.

7.4. COOPERATION WITH AMERICAN AND JAPANESE COMPANIES

While intra-European R&D cooperation agreements within the framework of ESPRIT, BRITE, RACE or EUREKA are sufficiently documented, it is far more difficult to get comprehensive information on the worldwide set of alliances in which the European firms participate. These encompass cooperation in R&D as well as equity agreements. In this section we will look at a number of prominent joint core technology projects of the European multinationals included in our sample. We will analyse the nature of these alliances and their relation to intra-European cooperation. We will attempt to answer the question whether or not European cooperation is a substitute for intercontinental cooperation.

7.4.1. The pattern of cooperation with
American and Japanese firms

The wide variety of cooperation agreements range from mere information exchange and licensing agreements to joint development or marketing contracts or a combination of these. Joint marketing in some cases results in outright cartelization. A distinction should be made between the different *goals* of cooperation, which could be exchanging resources or joint development, and the *nature* or form of cooperation, such as equity or non-equity based agreements.[16]

A growing number of studies[17] of multinational alliances agree that the non-equity type agreement is more prevalent than cooperation involving equity sharing characteristic of many joint ventures. This suggests that, in general, collaborating companies are eager to retain their independence *vis-à-vis* each other, despite the fact that they are increasingly eager to cooperate.[18] Research on cooperation agreements has also revealed that technological knowhow has become the most important factor behind cooperation agreements in the 1980s. The *Report on Competition Policy*, published annually by the European Commission, maintains that joint ventures since 1982 have become increasingly concentrated in high technology sectors.[19]

Table 7.5 lists the major international cooperation agreements in the core technologies during the period January 1980 to July 1986. 'Major' as opposed to 'minor' cooperation agreements have been selected for our sample on the criteria that (a) at least one of the European multinational firms contained in our sample is involved (with a couple of exceptions for a few relevant additional deals); (b) the agreement involves another large firm — Japanese, European or American; and (c) that the alliance involves a core technology area making it a 'strategic' alliance.[20] In Chapter Two we have shown that a large number of minor cooperation agreements and takeover deals are also relevant when analysing the strategy of specific firms. In this chapter we are less interested in individual strategies than the overall position of European multinationals in the international network of cooperation.

---

[16] See Gorbis and Yorke (1985).

[17] See Jacquemin (1986); Gabrielli (1986); Doz, Hamel and Prahalad (1986).

[18] Gabrielli (1986, Table 2), on the basis of his own databank, comes to the conclusion that 60 per cent of the around 1000 agreements observed in the 1982–1985 (first quarter) period are of a non-equity nature, whereas around 25 per cent involved joint ventures.

[19] Commission of the European Communities (1985e), p. 198.

[20] For more detailed overviews of cooperation deals see: Junne and Tulder (1984); Tulder and Junne (1984); Tulder and Empel (1984); ECE (1985, 1986); OECD (1985c, 1986a); Haklisch (1986).

Table 7.5 *Major international cooperation agreements involving European Multinationals in the 1980–6 period*

European firm		Nature of the agreement
Telecommunications		
PHILIPS	↔ AT&T (USA)	Joint venture in developing and marketing of switching exchanges outside the USA and SE Asia (1984)
PHILIPS (51%)	↔ Control Data (USA)	Joint venture ('Optical Storage International'); R&D marketing of optical peripherals equipment; in 1986 integrated in PHILIPS–DU PONT joint venture
PHILIPS	← Concord Data Systems (USA)	License of local area network technology
PHILIPS	↔ Kyocera (Jap.)	Joint venture, development of interactive telecommunications systems, marketing in Japan (1985) Joint venture to sell Philips Sopho wide area network in Japan (1985)
PHILIPS	↔ BOSCH	Joint venture for production, development and sales of television equipment 'Euro Television Systems' (1986) (50–50%)
THORN/ ERICSSON	→ United Technologies (USA)	Supply of mobile radio systems (NMT system)
GEC	↔ Mitsubishi (Jap.)	Marketing of satellite earth stations
GEC	← Northern Telecom (Can.)	Production under license of large PABX
GEC–MAROONI	← NEC (Jap.)	Production under licence of radio telephone equipment
PLESSEY	↔ Rolm (USA)	Digital phone technology (finished end 1983)
ICL	← Mitel (British Telecom)	Marketing of switching equipment (PBX)
BRITISH TELECOM/ BRITISH AEROSPACE	↔ IBM (USA)	Joint manufacturing (in UK) of TELECOM cashless shopping system
BRITISH TELECOM	← Du Pont (USA)	Opto-electrical components (1986)
BRITISH TELECOM	↔ Kokusai Denshin (Japan)	Joint venture digital networks
BRITISH TELECOM	↔ McDONNAL DOUGLAS (USA)	Joint venture in VAN
Telettra (FIAT)	↔ GTE (USA)	Set-up of subsidiary in USA for microwave systems for digital transmission (1986)
Telettra	↔ ANT	Joint venture cellular mobile radio
Telettra	← TIE (USA)	production of intercom systems in Italy (1984)

(continued)

Table 7.5   (continued)

European firm		Nature of the agreement
SIEMENS (80%)	↔ GTE (20%) (USA)	Joint venture (almost takeover) USA transmission activities and public switching in Europe (1986)
SIEMENS	↔ Xerox (USA)	Cooperation in PABX
OLIVETTI	← Intercom (USA)	PABX technology
OLIVETTI	← Northern Telecom (Can.)	Licensing PABX
OLIVETTI	← AT&T	Minimal acquisition of 25%
OLIVETTI	↔ Toshiba (Japan) (20%)	Joint venture for office automation, Japanese marketing of Olivetti equipment (1985)
ITALTEL	↔ GTE (USA)	Cooperation in digital switch development, production and export
CIT–ALCATEL	↔ PLESSEY/GEC/ ITALTEL	Cooperation in new generation of publication telephone exchanges
CIT–ALCATEL	↔ SIEMENS	Negotiations on cooperation in radiotelephones
CIT–ALCATEL	↔ PHILIPS	Cooperation in mobile telephone
THOMSON–CSF	← Xerox (USA)	Joint cooperation (Fortune Systems Corp.) for digital optical discs
ALCATEL– THOMSON	↔ Fairchild Industries (USA)	Joint activities in networks and services on each other's market (1985)
THOMSON	↔ Hughes/GM (USA)	Joint efforts in satellites
THOMSON–CSF	↔ PHILIPS	Joint development of common standard for a direct Broadcasting System
CIT–Alcatel	→ Honeywell (USA)	Marketing of videotex system
MATRA	↔ Tymshare (USA)	Joint efforts in videotex
MATRA	↔ Boeing Aerospace Company (USA)	Production and commercialization of small satellites
MATRA	↔ BRITISH AEROSPACE	Cooperation in satellites
MATRA	↔ BOSCH	Negotiations on cooperation in radiotelephones
BULL	↔ Trilogy Systems	Equity, technology development in components and peripherals
BULL	↔ Gen. Electr. (USA)	Cooperation in VANs for French market
PLESSEY	→ TELENOKIA (Fin.)	Supply of digital phone technology
PLESSEY	→ OLIVETTI	License of PABX
ERICSSON	↔ Honeywell (USA)	Marketing of Ericsson's system and joint R&D company in USA on PABX
ERICSSON	↔ THORN–EMI	Joint venture in UK

(continued)

Table 7.5   (continued)

European firm		Nature of the agreement
ERICSSON	→ THOMSON–CSF	Building under licence of digital switching system (AXE)
ITALTEL	↔ JEUMONT/ SCHNEIDER	Link up in PABX
BOSCH	↔ JEUMONT/ SCHNEIDER	R&D for networks and integration of services, mutual marketing in telecom (1985)
ITALTEL	↔ CIT–Alcatel	Joint venture in videotex

Robotics and automation systems

ASEA	→ Trallfa (Norway)	Joint venture, 51% participation (1985)
MATRA (Manurhin)	← Fujitsu Fanuc (Jap.)	Licences for manufacturing and sales
RENAULT	→ Ransburg (USA) (Cybotech)	Joint venture, 49%
RENAULT	← Toyoda (Toyota) (Jap.)	Joint venture, without shareholding, in R&D, technology exchange, manufacturing
RENAULT	↔ Federal Products Corp. (USA)	Joint venture, without shareholding, in R&D, technology exchange, manufacturing in coordinate measuring machines
OGE	← Sankyo Seiki (Jap.)	Sales
OGE	↔ Yaskawa Electric (Jap.)	Mutual licensing of industrial automation
OGE	← Toshiba (Jap.)	Marketing and sales agreement
OGE	← Osaka (Jap.)	Sales
THOMSON	← Dainichi Kiko	Marketing and sales agreement
VOLKSWAGEN	→ General Electric (USA)	Licences for marketing and sales
SIEMENS	↔ Fujitsu Fanuc (Jap.)	Joint venture without shareholding, in R&D, technological exchange, manufacturing
THORN–EMI	← Yaskawa (Jap.)	Marketing and sales agreement
SIEMENS	↔ KTM (UK)	Joint venture, software
GEC	→ Bink Bullows (USA)	Marketing and sales agreement
GEC	← Hitachi (Jap.)	Licences for marketing and sales
THORN–EMI	← Yaskawa Electric (Jap.)	Sales
STET/DEA	→ Amada (Jap.)	Licences for marketing and sales
STET/DEA	→ General Electric (USA)	Licences for marketing and sales

(continued)

Table 7.5 (continued)

European firm		Nature of the agreement
STET/DEA (Selenia–Elsag)	↔ IBM (USA)	Joint venture without shareholding in R&D, technological exchange and manufacturing (1985)
OLIVETTI	→ Westinghouse (USA)	Licences for marketing and sales
OLIVETTI	↔ Allen–Bradley (USA)	Joint venture, numerically-controlled systems (1982)
FIAT (Comau)	→ Bendix Corp. (USA)	Joint venture, 10% participation
FIAT (Comau)	↔ Digital (DEC)	Joint venture for CIM (1985)
ASEA	↔ Westinghouse (USA)	Cooperation
ASEA	↔ Kawasaki (Jap.)	Sales, research and development
HOOGOVENS (NL)	← Dainichi Kiko (Jap.)	Licence, sales
STET/DEA	→ Fairey (UK)	Licence
STET/DEA	↔ Kuka (BRD)	Technical cooperation

Semiconductor industry

PHILIPS	↔ Advanced Micro Divices (USA)	Technology exchange in bipolar microprocessors (1981)
PHILIPS	↔ RCA (USA)	Cooperation on ICs (1982)
PHILIPS	↔ Control Data Corp. (USA)	Two joint ventures in USA and the Netherlands (1982)
PHILIPS	↔ Intel (USA)	Technology exchange and cooperation in CHMOS development of ICs (1982)
PHILIPS	↔ Motorola	Second sourcing of 16 bit processors and development of new products also for Motorola (1981)
PHILIPS	↔ Matsushita (Jap.)	Joint venture in Japan with Matsushita Electr. Corp. (Philips holds 35%)
PHILIPS (40%)	↔ Nippon Steel (30%) Nippon Chemi-Con Corp. (30%)	Joint venture in Japan for manufacture and sale of fine ceramic electronics components (1986)
PHILIPS	↔ Texas Instruments (USA)	Second source and build-up of 'chip-library' (1984)
PHILIPS (Signetics)	↔ Texas Instruments (USA)	Technology exchange, development of OMOS ICs
PHILIPS	↔ Lineat Technology (USA)	Technical exchange agreement
PHILIPS	↔ BULL	Joint manufacturing of memory cards

(continued)

Table 7.5   (continued)

European firm		Nature of the agreement
PHILIPS	↔ Intel (USA)	Agreement on 16 bit micro controllers (1985)
SIEMENS	↔ West Digital (USA)	License chip technology
PHILIPS	↔ SIEMENS	Joint R&D on semiconductors, CAD and speech recognition (long term)
PHILIPS	↔ SIEMENS	Megabit project, cooperative research
SIEMENS	↔ Intel (USA)	Second sourcing, cooperation in ICs
SIEMENS	↔ Fuji (Jap.)	Joint venture in Japan for assembly of ICs
SIEMENS	↔ Toshiba (Jap.)	Joint development and production of ICs, submicron technology exchange (1985)
SIEMENS	↔ Gen. El. (USA)	Idem. (1986)
MATRA	↔ Harris (USA)	Joint venture for custom and specialized SCs (French financing, USA provides technology)
MATRA–HARRIS	↔ BULL	Special VLSI circuits using MOS technology
MATRA–HARRIS	↔ Intel (USA)	Joint venture to design micro-processors (Intel 49%) (1981)
MATRA–HARRIS	← Nec (Jap.)	Technical agreement proposed for 4 bit microprocessors
STET (SGS–Ates)	← Zilog	Second source for Z-series micro-processors (1980)
STET (SGS–Ates)	← Toshiba (Jap.)	Production and marketing of OMOS SCs and development of new ICs (1982); second source (1981)
STET (SGS–Ates)	↔ ERICSSON	Research, product development and possibly marketing collaboration (1984)
STET (SGS–Ates)	↔ AT&T	Marketing of AT&T chips (1986); joint venture chip circuit technology
THOMSON–CSF	← AMD, Gen. Instr., Gen. Electr. (USA)	Second-source agreements
THOMSON–CSF	← National Semi-conductors (USA)	Continued second sourcing after Thomson acquired NS share in a joint venture with Saint-Gobain (1983)
OLIVETTI	← Zilog	Microprocessors (1980)
GEC	← Mitel/British Telecom	Technology transfer of 'Cellmos' for internal use
GEC	← Intel (USA)	Second source for production of tele-communications components (1984)
FERRANTI (UK)	↔ GTE	Joint venture; Ferranti supplies ICs

(continued)

Table 7.5   (continued)

European firm		Nature of the agreement
BULL	← Trilogy (USA)	Licence agreement; access to Trilogy's wafer technology, licence for SC technology (1981)
ICI	← Fujitsu (Jap.)	Licence for SC technology (1981)
THORN–EMI/ Inmos	→ NMB Semi- conductors (Jap.)	Production under licence of 256 Dynamic RAM (1984)
SGS–Ates (STET)	↔ AT&T (USA)	Sales on US market (1986)
THOMSON	↔ SIEMENS, GEC, PHILIPS, PLESSEY	Design of a sophisticated processor for military and civilian applications (1985)

New materials

CIBA–GEIGY	↔ Ashai Chem. (Jap.)	Joint venture in carbon-fibre (1981)
PHILIPS	↔ Ma–Com (USA)	Joint venture in US fibre optics
PHILIPS	↔ Dupont (USA)	Joint venture for electro-optical information, compact disc (1985)
CIT–Alcatel	← Corning Glass (USA)	Production under licence of optical fibres
SAINT-GOBAIN/ THOMSON	↔ Corning Glass (USA)	Joint factory in France, Fibres Optiques Industries, 40% by Corning
PLESSEY	→ DSC Communication	Joint venture for glass fibre cables marketing, Plessey to have 15% of US market by 1990
PLESSEY	↔ Corning Glass	Cooperation in fibre optics
PIRELLI	← Corning Glass	Licence in fibre optics
BRITISH TELECOM	↔ Du Pont (USA)	Joint venture, laser-based communication through optical fibres
SIEMENS	↔ Corning Glass	Joint venture, Siecor, to produce fibre optic cable (1984)
Pechiney (F)	↔ Sumitomo (Jap.)	Joint manufacturing of carbon fibres
Enka Ga. (D)	↔ Nippon Carbon (Jap.)	Joint manufacturing of carbon fibres
DEGUSSA	↔ Sunnyvale (USA)	Production of ceramics for ICs
DSM	↔ Unitika (Jap.)	Research on fibres
DSM	↔ Toyobo (Jap.)	Development of superstrong polyetheleen fibre (1984)
FIAT	↔ VOLKSWAGEN, BL, PSA, RENAULT, VOLVO	To avoid duplication, by exchange, mainly in the field of new materials, to compete better with Japan and USA (1985)
HOECHST	↔ 3M (USA)	Joint production of fluor elastomers (high temperature resistant rubbers) (1985)

(continued)

Table 7.5 (continued)

European firm		Nature of the agreement
MONTEDISON	↔ Hercules (USA)	Joint venture (Himont), US based for polypropylen, fibres, composites and alloys (1983)
MONTEDISON	↔ Asahi Glass (Jap.)	Joint venture in Japan (1986)
AKZO	← Toho Rayon (Jap.)	License of carbon fibre
AKZO (majority)	↔ Kollmorgen	Joint venture in development and production of materials for electronics in USA (1986)

Biotechnology

Bio Gen (S)	→ Green Cross (USA)	Technological participation
Hoffmann–La Roche (CH)	← Genentech (USA)	Technological participation
Hoffmann–La Roche (CH)	↔ Takeda Pharmaceutical (Jap.)	Technological participation through joint subsidiary Nippon Roche
HOFFMANN–La Roche (Swi.)	← Immunex (USA)	Manufacturing and marketing agreements
DSM	↔ Toyo Soda (Jap.)	Joint venture in aspartame: Holland Sweetener Cie (1985)
SHELL	↔ Celltech (UK)	Joint research
SHELL	↔ Cetus (USA)	Joint research
SHELL	↔ GIST–BROCADES	Joint research
Kabi Vitrum (S)	← Sumitomo Chemical (Jap.)	Technological participation
Kabi Vitrum (S)	← Genentech (USA)	Technological participation
CIBA–GEIGY	← Bio-response (USA)	Manufacturing of cellular protein
CIBA–GEIGY	← Genentech (USA)	Licensing agreement for worldwide marketing and sales rights of gene-splicing veterinary products
NESTLÉ (Nestec)	↔ Calgene (USA)	Joint venture for 'pest resistent soja bean' (1985)
RHÔNE-POULENC	← Calgene (USA)	Research to develop new herbicide resistant varieties of sunflower
Roussel–Uclaf	← Cetus (USA)	Research on a vitamin B12 producing microorganism
Nordisk (DN)	→ Chiron (USA)	Research contract to develop rDNA derived factor VIII antihaemophilia and human insulin
Sandoz (D)	← Collaborative Research (USA)	Sales agreement

(continued)

Table 7.5   *(continued)*

European firm		Nature of the agreement
Boeringer Ingelheim (D)	← Genentech (USA)	Sales, some exclusive marketing rights for Genentech products outside USA
Boeringer Ingelheim (D)	← Molecular Genetics	Agreement to produce rDNA products
Kabi Gen (S)	← Genentech (USA)	Licences and sales
Pharmacia (S)	← Genex (USA)	Research agreement
Pharmacia (S)	↔ Chiron (USA)	Joint venture for therapeutic enzymes (1985)
Pierrel (I) (Fermentia)	↔ Genex (USA)	Research project and sales rights for Pierrel
Schering (D)	↔ Genex (USA)	Two research agreements for the development of rDNA products, process technologies and licensing agreements
Serono (I)	← Integrated	Research contract
Novo Industri (DM)	→ E.R. Squibb & Sons (USA)	Joint venture for marketing Novo's insulin in the USA
Kabi Vitrum (S)	← Genentech (USA)	Research agreement with Green Cross for production of serum using Genex rDNA technology

Computers and software

BULL	← Nippon Electric Co. (Jap.)	Technology development and marketing
BULL	↔ OLIVETTI	Joint development of banking automates cooperation, long term, technology exchange
BULL	↔ Microsoft (USA)	Cooperation in software (networks, graphical user interfaces) (1985/6)
SIEMENS (51%)	↔ STET (49%)	SIEMENS data (since 1960)
SIEMENS	← Fujitsu (Jap.)	Marketing agreement
SIEMENS	↔ Microsoft (USA)	Cooperation in software package
BASF	← Hitachi (Jap.)	Licences
OLIVETTI	← Hitachi (Jap.)	Licences
ICL	← Fujitsu (Jap.)	Licences and technological agreement, OEM of peripherals, after 1986 technology exchange
ICL	↔ SIEMENS, BULL	Agreement to exchange information between research centres; joint institute in Munich called European Computer Industry Research Centre (1984)
ICL	← Logica	OEM agreement in small computers

(continued)

Table 7.5   (continued)

European firm		Nature of the agreement
ICL	← PERQ SYSTMES	Development and manufacturing small computers
ICL	← RAIR	Manufacturing and marketing small computers
NIXDORF	↔ Pyramid Technology Corporation (USA)	Exchange of technology, joint development of 'supermini' computer (1986)
NIXDORF	↔ Spartacus Comp. (USA)	Technology exchange of licensing of software
NIXDORF	↔ Kenematsu Gosho	Joint subsidiary
NIXDORF	↔ Microsoft (USA)	Cooperation in PCs and micros; adaptation of software programmes (1985)
ERICSSON	↔ Sperry (USA)	Cooperation in banking automation
ERICSSON	↔ DEC (USA)	Joint company for development and production of banking information systems (1985/6) in Sweden
OLIVETTI	← Kyocera (Jap.)	OEM in small sized computers
OLIVETTI	← Corona	Marketing small computer
OLIVETTI	← Stratus Comp. (USA)	Equity, marketing (medium computers)
OLIVETTI	← AT&T (USA)	Medium computer marketing
Racal Electr.	↔ Norsk Data	Joint venture, artificial intelligence
PHILIPS	↔ DEC (USA)	Non-exclusive cooperation agreement in digital screen information (medical applications) (1985)
JEUMONT	↔ Wang (USA)	Agreement on technical collaboration (1984)
TELEFONICA	↔ Fujitsu (Jap.)	Joint ventures, Fujitsu Espana, to produce computers
BASF	← Hewlett–Packard (USA)	Technological agreement
SIEMENS	↔ Valid Logic Syst.	License of software (preferred European supplier) for CAE graphical systems (1985)
SIEMENS	↔ PHILIPS, BULL, ICL	Joint development of Unix software standard (1984/5)

In Table 7.5 cooperation in the following core technologies is listed: telecommunications equipment (47 alliances); robotics and automation systems (29); semiconductors (38); new materials (21); biotechnology (27); and computers and software (31). In total, 193 alliances are included.

It should be noted that we do not discuss how a takeover or a cooperation agreement between two firms can influence the pattern of cooperation with a third firm. An example here is the Rolm–Plessey deal in telecommunications equipment which was shelved after the firm was taken over by IBM. Also, it should be noted that the less equity involvement there is, the more the network becomes susceptible to rapid change.

No less than 59 per cent of all the international alliances in which European multinationals are involved are with US multinationals. In 23 per cent of the cases European multinationals have an alliance with an important Japanese firm. One third of these alliances are in the area of robotics. Only in 18 per cent of the cases did European firms start strategic alliances in core technologies with each other, that is on a bilateral and binational scale. In almost half of the cases these alliances were in telecommunications. This network of intra-European alliances more or less supplements the network of European cooperation projects discussed in the previous section.[21]

With regard to the *content* of the deals, roughly 60 per cent involved either information exchange, joint marketing or production in the form of a joint venture. The joint venture type of alliance represents almost half of such two-way relations and 27 per cent of the total sample.[22] In unilateral types of agreement in which technology (even in marketing agreements, but on a limited scale) is transferred from one continent to the other, European firms in three quarters of the cases are at the receiving end of the deal. In the area of biotechnology this type of agreement is used more frequently than the two-way exchange relation, which indicates that biotechnology is targeted by European multinationals as an area for technology transfer rather than of cooperation. Out of the total number of alliances the European multinationals have been suppliers of technology in only about 10 per cent. More than half of these unilateral arrangements have been in the field of robotics. Another major area is telecommunications, where firms seek to make inroads into the protected markets of other countries through licensing and technology transfer agreements.

In *joint ventures*, the most pervasive form of strategic alliance in which both firms give away part of their autonomy, we find the following pattern: in the period 1980–6 more than 50 per cent of all joint ventures of European firms (27) were with US firms and 31 per cent (16) were with Japanese firms. In the same timespan only 9 intra-European joint ventures

[21] The alliances under the EUREKA programme, especially, come close to the deals listed in Table 7.5; they have not been included in the table. This makes a comparative analysis on the basis of Table 7.5 more difficult.

[22] This figure corroborates the findings of Gabrielli (1986) who in a broader sample of alliances for a comparable period observed that 25 per cent of the agreements took the shape of joint ventures.

(17 per cent) involved core technologies. Joint ventures with American companies are concluded for a variety of reasons. In the case of the Euro–Japanese links a clear pattern emerges: 9 out of the total 16 joint ventures have been for gaining access to the Japanese market. At least 3 of the remaining agreements have been for an increasing market share in Europe. In the case of the Euro–Japanese joint ventures, at least three quarters aim at market entry, particularly to the Japanese market. The Japanese partners in these cases traded (European) market entrance to the Japanese market against technology transfer from Europe. A comparable deal of market entrance against technology transfer in joint ventures (especially towards American firms) has been used by French companies.

Marketing-directed joint ventures predominate in the Euro–Japanese corporate agreements. Most of the other relations are of a unilateral type. In only four instances has mutual cooperation among what might be called 'equals' been concluded: CGE–Yaskawa (robotics), DSM-Unitika (new materials), ASEA-KAWASAKI and SIEMENS-TOSHIBA. The two latter deals are examples of genuine triad arrangements involving robotics and chips in which ASEA and KAWASAKI cooperate on an equal footing with WESTINGHOUSE, and SIEMENS and TOSHIBA with GENERAL ELECTRIC.

This pattern of Euro–Japanese deals reveals that European multi-nationals follow a similar strategy in Japan as Japanese multinationals in Europe, which primarily is aimed at facilitating their internationalization process by using alliances to enter foreign markets which are, in many cases, protected. Once dealer networks or consumer loyalty is established, firms can more easily enter the market without any special agreements. Technology transfer in such agreements is very limited since the Japanese firms often use the European 'partner' as an assembly firm which is little more than a 'screwdriver factory'. International cooperation in such cases does not increase the technological capacities of the receiving firm and in the longer term may even reduce it. Of the twenty or so unilateral agreements with European firms at the receiving end, most are of a marketing nature. Hardly any substantial technology transfer is involved. At the same time we have found only two unilateral agreements between European firms and Japanese companies. In both cases the Japanese were able to tap the technology of a very innovative European firm which is not a major direct competitor. In the first case a licence was given by INMOS (THORN–EMI); in the second case robot technology was transferred by DEA (STET).

Finally, the intra-European exchange agreements are predominantly in the area of telecommunications, in basic research or in niches which depend on specific standards like mobile telephones. Cooperation in these cases (equity as well as non-equity based) often involves R&D and standardization arrangements intended to achieve compatibility with the ESPRIT and RACE networks.

Do European firms operate as junior partners? While in specific areas this may be true, the question is difficult to answer in any conclusive way because verification is difficult. In mainframe computers, for example, the European licensing partners of Hitachi, Fujitsu and NEC 'have become the front line of the Japanese manufacturers' challenge against IBM' and, according to Doz, Hamel and Prahalad, 'through this process, the [European] partners have become cannon fodder in a global fight, and may, ultimately, be worse off than they would have been before the partnership'.[23] It remains to be seen, however, whether the alliances of SIEMENS with Fujitsu, of BULL with NEC and of OLIVETTI with Hitachi have not served the European firms well in complementing their product range in a pragmatic way. It is remarkable, for example, that in the triad relationship in computers between NEC, BULL and Honeywell, in 1987 the American firm finally handed over a majority share to BULL which indicates that the French firm may not have been such a weak partner in the coalition after all. Many European firms, however, do seem to function as junior partners in the case of the alliances in robotics with Japanese firms. 'Second-sourcing' agreements with US semiconductor firms (see Chapter Two) have *severely limited the ability of the European partner* to sell in the markets of its American partner. This is because the contracts usually only apply to European markets and in many cases prohibit the European producer from selling the product (or he product in which a particular component is incorporated) in the US market. This can make it difficult for European producers to achieve scale economies. This is particularly true in the case of chips, in which Europe's world market share is only 18 per cent, and that of individual European countries like France only amounts to 3 per cent of the world market. The numerous second-sourcing agreements THOMSON, for example, has with American firms in semiconductors conflicts with its explicit policy of increasing its share of the American market.[24] Another area in which European firms increasingly tend to function as a 'junior partner', which is not included in our sample, is computer aided design.

Are European and global cooperation agreements in *conflict* with each other? A specific example of cooperation between an American and a European firm which affected the latter's European cooperation strategy has been the joint venture in 1985 between the STET robotics subsidiary DEA and IBM. This prompted the Italian firm to withdraw its subsidiary from the CIM project of ESPRIT. In explaining its decision STET argued

[23] Doz, Hamel and Prahalad (1986).
[24] See *Electronic Times*, 30 May 1985, p. 11.

that it could not finance both the European cooperation project and cooperation with IBM.[25] The Italian firm set its priorities in favour of its transatlantic link.

Often, European firms use their alliances with US or Japanese companies as a *substitute* for cooperation with other European firms. The cooperation agreements are used as *competition by other means*. In that sense European and global agreements indeed contradict each other!

Do the different networks *supplement* each other? The bulk of the European cooperation networks are in research and development. Most of the international cooperation agreements are closer to, or are aimed at, direct commercialization. In this sense the networks supplement each other in the different phases from basic research, through development, to commercialization and marketing. Since the European networks predominantly only aim at the basic research phase, European cooperation programmes can be regarded as a *precondition* for many intercontinental cooperation programmes. This has been suggested as one of the major aims of French national cooperation programmes. Chesnais[26] observes that French Government support for cooperation programmes was aimed less at national independence than in order to appear as an attractive partner for international cooperation and not *run the risk of being excluded from international technology transfer*. This has also been one of the main reasons behind some of the national British programmes, and might easily be extended to the European level. Especially when progress in technology comes from military developments this will be the case. Firms which are shut out of major programmes, and which cannot hope to be able to buy a particular patent or licence in a strategic technology because the country's ministry of defence will prohibit this, have to find an alternative means to establish strategic partnerships with the firms involved in these military programmes. It is possible that this kind of reasoning has also been behind the French EUREKA initiative as a reaction to the American SDI programme. The SDI programme is not expected to result in major cooperation agreements with European firms.

7.4.2. European cooperation: no substitute for intercontinental cooperation

Section 7.3 on European networks of cooperation showed that the intensity of European cooperation has increased considerably during the 1984–7

---

[25] Interviews.
[26] Chesnais and la Saussay (1985).

period. In section 7.4 it has been shown that intercontinental cooperation often serves as a substitute for European cooperation. Now this problem will be reversed and the question raised whether Europe-wide cooperation can take the place of closer cooperation with American and Japanese firms and whether intra-European and transatlantic cooperation agreements complement each other—that is, cannot function as an alternative for the other. To find an answer to this question, we shall turn to the different reasons for international cooperation agreements as described in section 7.2. We shall see whether or not the different objectives of cooperation agreements can be attained by agreements within Europe, or whether certain objectives of such agreements can only be realized by agreements with firms from the United States and Japan.

Risk sharing   The risk that a competitor will bring a specific product to market more quickly should obviously be shared with the company that is the most advanced competitor, whether this is an American, a Japanese or a European company. In many cases the most important competitor will come from the United States or Japan. In this case cooperation with a European partner would not significantly reduce the risk of another company being first in the market.

Risk of market access   Since the problem of market access within Europe only plays a role in so far as government procurement is concerned, the inclination to cooperate with companies from other European countries to acquire better market access is much less than in the case of an agreement with an American or Japanese firm. In the Japanese market few European companies can go it alone without a Japanese partner. To get access to additional markets within the OECD regions, therefore, can be an important stimulus for intercontinental cooperation which cannot be substituted by cooperation within Europe. In the case of access to the significantly smaller markets of Third World countries, intra-European cooperation is more likely to occur. This has been the case, for example, with the PHILIPS–ERICSSON consortium for providing Saudi Arabia with telecommunications equipment.

Risk of market acceptance   Consumption patterns tend to be similar in the countries of the Western hemisphere, i.e. in Europe and the United States in comparison, for example, with consumption patterns in Japan. Cooperation with a Japanese partner, therefore, may be necessary to adapt a product to Japanese taste and consumer preferences. Where this is the case, it is obvious that cooperation with a European partner could not fulfil the same function.

Manifold applications of core technologies If a European company has developed a technology that can also be applied to a field outside its own range of activities, it can look for a European as well as a non-European partner. It is probable that the preference will go to the company that has the strongest market position in the area in question, because the latter will be able to pay the highest price for the technology offered. But cooperation with a European firm in this case can be a real alternative to cooperation with an American or Japanese firm. If a company, however, needs to procure a technology developed in another field, which is also applicable to its own activity, it will have to cooperate with whoever developed the particular technology in question (for instance, with Corning Glass). If it is an American or a Japanese company, an agreement with a European partner cannot be a substitute. In new materials and biotechnology European cooperation is in a broad range of cases quite feasible. In the microelectronics area, Japanese and American companies dominate specific parts of the cluster. The leading role that Japanese companies play in a number of areas related to production automation (for instance, in chip production technology) makes Japan the most promising avenue for cooperation for a European company. The SIEMENS–PHILIPS Megachip project, for example, was supplemented by a SIEMENS–TOSHIBA agreement in order to keep track of the most advanced production technology which is an indispensable input for further progress in this area.

Complementary developments in different technology clusters To get access to expertise in other basic technologies, a company may cooperate with another company which possesses this technology. In most cases, a company of the same national origin will be chosen in order to keep communication barriers to a minimum. Only if the expertise sought does not exist in the own country will non-national cooperation partners, including those on other continents, be considered. If potential partners exist in Japan or in the United States that have far more experience than candidates for cooperation in Europe, then again cooperation with partners within Europe may not be a viable alternative.

The need for systems technology Any system becomes more attractive the more advanced its components are. A company that wants to offer complete systems, therefore, will be inclined to look for the most advanced and/or cheapest producers of the individual items, irrespective of their European, Japanese or American origin. Partners from Japan or the USA might even be preferred since they may be more willing to market the system components of a European partner than of another European company. Marketing products of another European firm would only create an additional competitor in a firm's home market, with little or no chance of enlarging the latter.

Establishment of dominant standards   Most standards actually have to be *world* standards. Cooperation with an American and/or a Japanese firm probably is more effective in establishing a world standard than cooperation with a European partner. On the other hand, it may be a firm's intention to develop a 'counter-standard' in order to reserve a specific market and protect it against the penetration of a potentially dominant outside competitor. In such a case, cooperation with as many other European producers as possible would be necessary.

Given the above arguments, it is clear that in the majority of cases cooperation with a European partner is *not* an alternative to cooperation with US or Japanese partner. Cooperation with a European partner in many cases simply would not help to realize the desired goals. Even if Europe were the home-base of as large a number of highly competitive companies as the United States or Japan, the more suitable cooperation partner in about two out of three cases would be extra-European! It is therefore unjustified to perceive intra-European agreements as an alternative to intercontinental cooperation.

7.5. CONCLUSION

The development of new technologies has been an important push factor behind the wave of new international cooperation agreements between companies that have been concluded in the first half of the 1980s. To find the right international partners has become a central strategic issue for most firms, an issue which is as important as the level and direction of spending on research and development. Alongside the subsidy race and the deregulation race referred to in earlier chapters, there is also a kind of cooperation race, in which firms scramble for the most attractive foreign cooperation partners.

European firms seem to have been somewhat less engaged in this process than American and Japanese firms. But this does not necessarily mean that they lag behind. They may not feel the need as much as do American or Japanese firms. Since they are more diversified than American firms, they can often turn to another division inside their own company to solve a specific problem. They do not face the same problems as Japanese concerns with regard to market access (with the exception of the Japanese market), and therefore do not have to conclude agreements that facilitate the penetration of foreign markets. And they do not have to conclude many bilateral agreements with other European firms, since the programmes of the European Community provide them with channels for information exchange and coordination of their research efforts.

European research programmes got off the ground in the early 1980s, because the possibilities for a concentration of national efforts had become exhausted, some national programmes turned out to be counterproductive for the development of a Common Market, and the competitive position of the European Community had worsened, especially in the field of microelectronics. This created a situation in which twelve European multinationals could launch the ESPRIT programme which had largely been written by their representatives. The apparent success of ESPRIT sparked off similar efforts in other fields. A comparable programme in biotechnology, however could not be achieved, because interests of European multinationals were much more heterogeneous in this field, their competitive position is less threatened, and the budgetary constraints of the European Community have meanwhile become tighter.

A very delicate problem has been the participation of European subsidiaries of foreign companies in European programmes. This applies almost exclusively to subsidiaries of American companies, because subsidiaries of Japanese companies hardly carry out any research in Europe yet. Formally, there are no barriers to the participation of subsidiaries of non-European companies. And indeed, subsidiaries of American firms do participate. IBM takes part in different programmes. Given the explicit purpose of the European programmes to strengthen Europe's position *vis-à-vis* American and Japanese competitors, however, the participation of foreign subsidiaries has informally been kept limited. It may also have been an act of self-restraint that IBM only submitted proposals to the CIM part of the ESPRIT programme. IBM probably had an interest in avoiding an ardent discussion on its position in Europe, which would have certainly taken place if the company had tried harder to participate in more programmes.

While a precarious compromise could be reached to restrict direct participation of European subsidiaries of non-European companies, the *indirect* participation of non-European companies cannot be excluded. Most European participants in the European projects have also entered into cooperation agreements with American and Japanese companies, which in this way can profit from the European research efforts. These cooperation agreements, however, normally do not involve a one-way flow of information (and if they do, they are mainly *towards* Europe and not out of Europe.) For the information that European companies release, they get something back. But in order to conclude such reciprocal arrangements, European companies have to boost their own research performance, and Europe-wide cooperation could help to do so. In many respects, the European programmes thus can help European companies to acquire the 'entrance ticket' for cooperation agreements with non-European companies.

That European companies are inclined to conclude more bilateral cooperation agreements with non-European than with European firms may be due to the fact, paradoxical as it sounds, that they are still very much European. That means that most of their sales come from European countries. Other European companies, as a consequence, are their main competitors. In order to improve their position against these competitors, they prefer to cooperate with non-European companies whose markets show less overlap. The big exception, of course, is the electronics industry in general and the computer industry specifically, where the most formidable competitor of all European companies is IBM. The competitive advantage of IBM is so big, however, that cooperation among European companies alone would not be sufficient to reach a strong position. Only cooperation with another American company with an equally strong technology base (AT&T), or Japanese companies, provides a solution in this situation.

This is one of many reasons why cooperation with other European companies cannot be regarded as a possible substitute for cooperation with American or Japanese companies. From a European perspective, it might seem at first sight that cooperation agreements among companies from different European countries should be given preference above cooperation with non-European companies. Such a defensive move, however, may in the long run not only strengthen but also weaken European competitiveness, where this leads to a further monopolization of the European market, a protectionist attitude towards non-European competitors, and a lack of expansion into markets outside Europe. Cooperation with non-European partners which implies a one-sided penetration of the European market, permanent status of the European firm as a junior partner, and eventually even the decline of the European potential for further research and development, however, would be equally bad. A choice between cooperation among European firms and cooperation with an American or Japanese firm therefore cannot be made in the abstract. The concrete situation and the terms of the respective agreement have to be studied carefully before any judgement can be given on the probable long-term impact of the resulting pattern of cooperation.

Conclusion

This book started from the assumption that it is not very useful to distinguish between 'high tech' sectors and non-'high tech' sectors, because *all* sectors are susceptible to the application of new technologies. For the business manager, for the trade unionist, for the politician or civil servant, as well as for the academic researcher, it is more fruitful to use the concept of *core technologies*, which we defined as follows. Core technologies (a) lead to a large number of new products and production processes, (b) are applicable across a large number of economic sectors, and (c) make a contribution to overcoming the obstacles created by the fast postwar accumulation. That is, they save labour and capital and provide more flexibility in their use, economize on energy and raw materials and cause less pollution.

We identified three clusters of core technologies: *microelectronics* (with its applications in semiconductors, telecommunications, robotics, computers, CAD equipment and software), *biotechnology* (the application of gene technology, enzyme technology and bioprocess engineering) and *new materials*. New materials received somewhat less attention than they deserve, because their range of applications is so diverse that they would have been impossible to deal with adequately without going beyond the scope of this volume.

Developments related to the three clusters are closely linked to one another, particularly in microelectronics and biotechnology. But the linkages between all three clusters are increasing. Advances in one cluster often become a precondition of further progress in another. The result is a complex, changing configuration of technologies which companies have to master in order to advance in their own field. This is part of the explanation of the increasingly intense network of international cooperation agreements.

European multinationals have made considerable efforts to expand their activities in the field of core technologies. Many of them have become active across the whole range of core technologies in a given cluster, and many have even expanded into several clusters (see Table 2.5). This is no longer a traditional form of commercial diversification which aims either at a wider spread of risks or at the penetration of other sectors with a higher profitability. Instead, companies become active in these fields in order to secure strategic inputs for their main activity. These strategic inputs are often not material components, but technologies or system knowhow.

The fact that many different companies from different economic sectors and industrial branches enter the same fields has led to an intensification of competition. Sector boundaries have become blurred, since all branches increasingly rely on the same basic technologies. This new competition leads to a 'reshuffling of the pack' even in branches with an established oligopolistic market structure (such as telecommunications). The intensified competition as a result of the increasing 'permeability' of sector boundaries, together with the additional competition as a result of increasing internationalization, has led to an assault on antitrust regulations at the national and international (European) level. This is justified with the argument that international competition has stiffened to such an extent that the fear of a too high concentration of market power has become obsolete. In the interests of international competitiveness, a higher concentration at the national level should be allowed, therefore. In many national economies (and at the level of the European Community) regulations have consequently been loosened.

The current increase in competition may not be permanent, however. The trend towards greater specialization and cooperation will necessarily lead to a higher degree of concentration. This concentration process will cut across sectors much more acutely than in the past. The recent expansion of DAIMLER–BENZ into electronics and aviation is an important example of this trend. This suggests that antitrust regulations may at some point have to be redressed in order to limit and control the market power gained by a number of corporations and corporate alliances. It is easier, however, to keep these regulations untouched than to return to them later: the 'damage' will have occurred, and with the newly formed larger corporations, formidable opponents of antitrust regulations will have been created that may effectively block the way to any more rigorous legislation.

The social impact of the application of new technologies by the European multinationals as well as by allied smaller companies will be far-reaching. Multinational companies appear to be concentrating increasingly on core technologies, produced and developed by a core workforce in core production facilities, located in core regions of core countries. This will obviously increase the segmentation of the labour market, augment regional imbalances and, thus, contribute to creating a considerable potential for social conflict in the future.

The effects on employment and on the quality of work will be felt differently by different categories of workers. This is an important reason why there is little organized opposition to the application of new technologies. The workforce is split into some groups which can expect to reap advantages from their introduction and other groups which will have to bear the costs. The two most negatively affected categories will be unskilled workers and middle management. It is often the opposition from this last group rather than from the shop floor which protracts the introduction of new technologies. New monitoring and control technologies effectively alter the power structure in affected companies. The degree and speed of the introduction of new technologies, therefore, are also the result of an internal power struggle in the companies. While the opposition of workers who become redundant may be more palpable and vocal and may spark more spectacular action, it may prove ineffective due to a lack of power to oppose the introduction of new technologies. These workers are played off against the 'core' skilled workers. The latter are in fact interested in further rationalization because it could increase their job security. The 'silent opposition' from middle management may prove more effective because it often obstructs the introduction of new management tools without open opposition.

Since most trade unions organize skilled workers rather than unskilled, most unions have taken a rather constructive stand on new technologies. Strong trade unions and a consensus climate between unions and employers has as a result often increased international competitiveness of a national economy. While trade union opposition to new technologies is often branded a major stumbling block in the public discussion, it is hardly mentioned by European managers as a factor impeding rationalization in their own companies. A lack of personnel with specific technical expertise is generally regarded as the most important obstacle. What appears to be a lack of expertise may often be the passive resistance of middle management.

Trade unions are as aware as other economic and political sectors that the distribution of the social consequences of the introduction of new technologies depends very much on the international competitiveness of individual companies and whole economies. In societies in which many new products are generated for the world market, new jobs are created. The same is true for economies which produce equipment used in the rationalization process. While there are a few countries or regions within countries where the overall balance is positive, there are many examples of countries or regions in which the negative impact markedly increases due to the fact that the more successful countries export unemployment. These economies cannot serve as a model, because it is impossible that all countries specialize in the production of rationalization equipment.

This 'zero-sum' character of the distribution of the negative social consequences of the introduction of new technologies explains in part the trend towards stifling international competition. Not only firms but whole societies are pitted against one another ('Japan Inc.' against 'Europe Ltd').

Countries are competing in a 'restructuring race' in order to provide the best environment for the expansion of their 'national' capital and to attract foreign investors. As in any race, it is not enough to run fast. What is important is to run faster than the others — or a least not to stay too far behind.

In an analysis of international competitiveness, it is not sufficient to analyse the present competitive position of the main companies from the major advanced industrial regions, the United States, Japan and Western Europe. Their respective positions can change quickly. Given the dialectics of progress, the advantage of any given historical moment can be a disadvantage in the next. What is important, therefore, is an analysis of the factors that allow companies or whole societies to bounce back after a temporary setback. SIEMENS, for instance, came to the market with a very sophisticated electro-*mechanical* telecommunications switching system — at a time when other companies were developing fully electronic systems. Many people believed that SIEMENS was out of the race and would not become a serious competitor again. They failed to consider SIEMEN's knowhow and experience which in fact allowed it to build sophisticated systems. With this in mind, we not only looked at the market position of firms in different high technology markets, but also examined various institutional and sociological factors which tend to change less rapidly.

Such factors include a firm's financial structure and the national financial system; the educational system and the traditional cultural values of a society embodied in its educational institutions; the relationship between the educational system (especially higher education) and industry, agriculture and services; the industrial structure of a country; existing networks of cooperation between small and large companies as well as among larger companies. A closer look at these factors often reveals that differences inside Europe are even larger than those between a European average and the United States and Japan. In a few countries (notably Sweden, the Federal Republic of Germany and Switzerland), the institutional preconditions of innovation, restructuring and accumulation are more like those in the United States and Japan. The differences are more acute, however, in the case of France and Great Britain. Such institutional factors merit more attention in studies of international competitiveness — even more attention than they have received in this book. The fact that institutional factors in a firm's country of origin have important effects on multinational enterprises underlines the limited extent to which most of these companies are truly 'multinational'.

In spite of the large differences within Europe, there are nevertheless several general characteristics of the large European enterprises. One is the greater diversification in general than their American counterparts.

This brings with it, however, the disadvantage that research and development as well as investment tend to be stretched over a broader range of sectors, decreasing the chance to mobilize a 'critical mass' of resources in any one given field. European companies often encounter difficulties in adapting their structure to new priorities because of the entrenched positions of their different divisions. This is made worse by the comparatively low mobility of middle management, whose loyalty often goes more to the division than to the company as a whole. This low mobility has two negative consequences for international competitiveness. One is the potential for more rapid diffusion of new technologies, which theoretically could be achieved by European companies precisely because of their more diversified structure. But such a diffusion is not realized because one of the most effective means of technology transfer — mobility of people — is not properly utilized. The other disadvantage is that the identification with a specific unit inside the company is relatively strong, which makes it difficult to mobilize resources for new priorities.

The missing resources then often have to be mobilized by government. Governments have become heavily involved in the support of the development and diffusion of new technologies. It has been explained that they are not only caught in a subsidy race, but also in a deregulation race. Whether direct and indirect subsidies or deregulation is emphasized depends on a country's size and its position in the international division of labour. The fact that deregulation plays such a big role in Great Britain, for example, is partly due to the fact that the country's industrial base has deteriorated to such a degree that it has become imperative to attract American and Japanese investment.

That countries are perceived as competing with each other in an international race objectively plays into the hands of the largest companies. This is because it helps to channel more support to the 'national champions', which are often depicted as representing the particular country in this race. In most European countries, R&D subsidies are highly concentrated and go to the largest companies. Even if governments tried to stimulate small and medium sized companies more effectively, they would probably only be able to implement a marginal shift in the subsidy volume. This is because the whole institutional structure is geared towards providing large subsidies to large companies. At the same time, however, this can put the international competitiveness of an economy at additional risk, because it may leave large sectors of the economy (where no 'national champions' exist) open to extensive penetration by foreign companies.

The strong role of national governments in developing and applying new technologies has not only advanced the competitiveness of European countries, it has also the negative side-effect of increasing non-tariff barriers within the European Community. This reduces the market for the most successful European as well as American and Japanese companies. Recognition of such counterproductive effects of national policies combined with considerable pressure from the larger European companies has initially led to the increased role of the European Community in developing new technologies. Having only a small budget, the Community tries to make a virtue of necessity by concentrating on stimulating cooperation among companies and research institutes in different European countries rather than spending large sums itself. This support intends to foster Europe-wide cooperation as an alternative to cooperation with American and Japanese companies, in which European firms often end up as a junior partner and stand to lose more in the long run than they gain. European companies, however, often perceive cooperation with other European companies from a different angle. They regard it as a chance to improve their bargaining position *vis-à-vis* American and Japanese partners. This helps them to become or remain an attractive international cooperation partner. From the company point of view, different cooperation agreements do not offset but rather complement each other. While most of the projects organized by the European Community aim at cooperation in pre-competitive research, cooperation with American and Japanese companies more often concerns areas nearer to commercialization. Given the major reasons for international cooperation (risk sharing, market access, commercialization of research results, extension of the technology base, establishment of common standards), cooperation with another European firm often is not a real alternative for cooperation with an American or Japanese firm. It is, therefore, no contradiction that the same companies whose executives are most actively engaged in promoting a common European technology policy, cooperate closely with American concerns. The examples of PHILIPS and OLIVETTI show that the motivation of European companies to formulate European programmes is not necessarily reduced as a result of bilateral cooperation agreements with firms outside Europe.

Still, the effectiveness of European technology policy to increase the competitiveness of European companies *vis-à-vis* American and Japanese companies may be negatively affected because common standards and regulations would be to the benefit of the outside cooperation partner as well; European government procurement and infrastructure investment would imply (at least indirect) orders to companies from the United States and Japan, and the knowledge accumulated with the help of subsidized research and development would leak away to cooperation partners from outside Europe. While the largest companies can permit themselves to engage in European cooperation projects as well as in projects with American and Japanese companies, smaller companies may lack the financial and management resources to pursue both at the same time.

Some tension will continue to exist, therefore, between the expansion of cooperation within Europe and increasingly intensive cooperation with American and Japanese firms. The interest of European multinationals in common European programmes, however, may be less undermined by cooperation with companies from outside Europe than by their own expansion into other regions. The rapidly increasing investment of European multinationals, especially in the United States, where they also expand their research base, may contribute to a shift of focus away from the European context. The strong support by European multinationals, especially from the electronics industry, for common European technology programmes may turn out to have been bound to a specific conjunctural situation which no longer exists, as the protracted tug-of-war for the medium-term European technology programme in 1987 illustrates.

The groups of European multinationals that were formed in the early 1980s to speed up the process of European integration had a two-fold purpose. They were not only lobby groups to influence the shape of programmes of the European Commission. They served at the same time as an instrument and a communication channel for the Commission to exert some pressure on the governments of member countries. This influence no longer seems to be used in the same way as in the early 1980s, at least in the three largest member countries (from which nine out of the twelve most important electronics multinationals come). With the EUREKA programme, an alternative framework has been created for more intensive cooperation between companies that wish to take part, outside the technology programmes of the European Community. In this programme the smaller European countries and smaller companies have even less influence than in the Community programmes.

Much more than in the United States, core technologies in Europe are developed and applied by multinational corporations first of all. They give shape to the direction of development, and they select the areas of application. Core technologies in principle can make a crucial contribution to overcoming the obstacles into which the postwar accumulation process has run (see Chapter One). But the concentration of control over these technologies seems to create other barriers to further accumulation. These include the increase of unemployment, increasing concentration of control over formerly independent companies, tendencies towards a dual society and the threat of increasing peripheralization for part of Europe (see Chapters Three and Four). If not counteracted by other actors, the beneficial potential of the new core technologies may remain underutilized.

Bibliography

Aglietta, Michel (1976), *Régulation et crises du capitalisme. L'Expérience des États-Unis*, Paris: Calman-Levy.

Altshuler, Alan, Anderson, Martin, Jones, Daniel, Roos, Daniel, and Womack, James (1984), *The Future of the Automobile. The Report of MIT's International Automobile Program*, London, Sydney: George Allen & Unwin, 320 pp.

AMES/AFRI (Automatisation et Mutations Économique et Sociales/Association Française de Robotique Industrielle) (1984), *Colloque organisé sous la patronage du Programme Mobilisateur 'Technologie, Emploi, Travail'*, Paris, 257 pp.

Anastassopoulos, Jean-Pierre, Blanc, Georges and Dussauge, Pierre (1985), *Les Multinationales publiques*, Genève: PUF/IRM, 256 pp.

Andries, Mireille (1985), *The EEC and Information Technologies*, Brussels: Agence Européenne d'Informations, 106 pp., annex.

Arnold, Erik (1984), *Industrial Adjustment and Policy: Computer Aided Design*, European Paper, no. 14, Sussex European Research Centre, University of Sussex.

Arnold, Erik, and Guy, Ken (1986), *Parallel Convergence. National Strategies in Information Technology*, London: Frances Pinter, 220 pp.

Bayer AG (1985), *Menschengerechte Gestaltung der Arbeit an Bildschirmgeräten*, 43 pp.

Berendsen, H. J. C. (1985), The relation between biotechnology and information technology, in European Parliament, Committee on Energy, Research and Technology, *Biotechnology Hearing (Outline)*, November, PE 98.227/rev.

Bessant, John (1984), *Flexible Manufacturing Systems. An Overview*, Geneva: United Nations Industrial Development Organization, Microelectronics Monitor, Supplement to no. 12.

BMFT (Bundesministerium für Forschung und Technologie) (1984a), *Informationstechnik. Konzeption der Bundesregierung zur Förderrung der Entwicklung der Mikroelektronik, der Informations- und Kommunikationstechniken*, Bonn, 80 pp.

BMFT (1984b), *Bundesbericht Forschung*, Summary, Bonn.

BMFT (1985), Programm Bioinformatik, Bonn.

Bolwijn, P. T. and Brinkman, S. (1985), *Verdelen en Beheersen. Japanse Fabrieken*, Eindhoven: Philips TEO, 100 pp.

Boyer, Robert and Petit, Pascal (1980), Emploi et productivité dans la CEE, *Economies et Statistiques*, no. 121, Avril–Mai.

Bull, Alan T., Holt, Geoffrey and Lilly, Malcolm D. (1982), *Biotechnology. International Trends and Perspectives*, Paris, OECD.

Bundesministerium der Finanzen (1985), *Zehnter Subventionsbericht*, Bonn, 323 pp.

Bundesministerium für Wirtschaft (1985), *Studien Reihe 47*, Bonn.

Buttel, Frederick H., Cowan, J. Tadlock, Kenney, Martin and Kloppenburg, Jack, Jr. (1984), Biotechnology in agriculture: the political economy of agribusiness reorganization and industry-university relationships, *Research in rural sociology and development*, 1, pp. 315–48.

Bylinsky, Gene (1983), The race to the automated factory, *Fortune*, February 21, pp. 52–64.

Carlsson, Jan (1980), The Swedish computer and electronic industry, structure and policies, Stockholm, mimeo.

Carlsson, Jan (1983), *Production and Use of Industrial Robots in Sweden in 1982*, Computers and Electronics Commission, DS I 1983: 11, pp. 1–44.

Cawson, Alan, Holmes, Peter and Stevens, Anne (1985), The interaction between firms and the state in France. The telecommunications and consumer electronics sectors, Cambridge, paper prepared for a conference on 'Government–Industry Relations in the Major OECD Countries', Trinity Hall, 10–13 December, 42 pp.

Chanaron, J. J., and Banville, E. de (1985), *Le Système automobile francais. de la sous-traitance au partenariat? Eléments d'une problématique*, Paris: Centre de Prospective et d'Evolution, no. 56, 77 pp.

Chenard, A., Pino, A. (1984), *L'Emploi dans la filière robotique*, Paris. Programme Automatisation et mutations économiques et sociales, Centre de Recherche en Economie Industrielle, Université Paris-Nord, October, 220 pp.

Chesnais, François, and la Saussay, Philippe de (1985), Some notes on French policy towards the acquisition of foreign technology, Ottawa: Six Countries Programme, draft of paper presented at the workshop on 'International Technology Transfer: Promotion and Barriers', 35 pp.

Ciborra, Claudio (1986), *Le Affinità asimmetriche. Il Caso Olivetti-AT&T*, Milan: France Angeli Libri s.r.l., 299 pp.

Cohen, Elie, and Bauer, Michel (1985), *Les Grandes Manoevres industrielles*, Paris: Belfont, 303 pp.

Cohendet, Patrick (direction) (1984), *La Chimie en Europe. Innovations, matations et perspectives*, Paris: Economica, 313 pp.

Commissie Steenbergen (1985), *Signalen voor Straks. Een nieuwe richting voor de PTT*, The Hague: State Publications Office, 50 pp.

Commission of the European Communities (1983a), *Proposal for a Council Decision Adopting the First European Strategic Programme for Research and Development in Information Technology (ESPRIT)*, Brussels, COM (83) 258 final.

Commission of the European Communities (1983b), *Biotechnology: The Community's Role, 'Background note': National Initiatives for the Support of Biotechnology*, Brussels, COM (83) 328 final/2.

Commission of the European Communities (1984), *A Scientific and Technical Strategy for Europe—Framework Programme 1984-1987*, Brussels, 99 pp., annex.

Commission of the European Communities (1985a), *new Technologies and Social Consequences*, Brussels: Directorate General for Research and Documentation, 18 pp.

Commission of the European Communities (1985b), *Towards a European Technology Community*, Memorandum, Brussels, COM (85) 350 final.

Commission of the European Communities (1985c), *Venture Capital in Europe 1985. Survey on Venture Capital in the European Community*, Brussels, Report EUR 10224 and Peat Marwick Mitchell & Co., 93 pp.

Commission of the European Communities (1985d), Advanced manufacturing equipment in the community, Brussels, Communication from the Commission to the Council.

Commission of the European Communities (1985e), *Fourteenth Report on Competition Policy*, Brussels.

Coombs, R. W. (1984), Long-term trends in automation, in Pauline Marstrand (1984), *New Technology and the Future of Work and Skills*, London: Frances Pinter, pp. 146–62.

Coriat, Benjamin (1984), *La Robotique*, Paris: Découverte, 125 pp.

CPE (Centre de Prospective et d'Évolution) (1985), *Grappes technologiques et stratégies industrielles*, Paris, 210 pp.

Dang Nguyen, Godefroy (1985), Telecommunications: a challenge to the old order, in Margaret Sharp (ed.) (1985a), pp. 87–133.

Dankbaar, Ben (1986), *Nieuwe Technologieën op de werkplek, ervaringen met het Westduitse programma 'Humanisering van de Arbeid'*, Amsterdam: TPE, March, no. 3.

Davidson, William H. (1984), *The Amazing Race, Winning the Technorivalry with Japan*, New York/London: John Wiley, 270 pp.

Delapierre, Michel, and Zimmermann, Jean-Benoit (1984), Les Multinationales de l'électronique: des stratégies différenciées, in *Revue d'économie industrielle*, no. 28, 2ème trimestre, pp. 9–36.

Dosi, Giovanni (1981), *Technical Change and Survival: Europe's Semiconductor Industry*, Sussex European Papers, Sussex European Research Centre, no. 9, 103 pp.

Doz, Yvez, Hamel, Gary and Prahalad, C. (1986), International strategic partnerships. Paper written for the Strategic Management Society. Quoted in the *Financial Times*, 17 October 1986.

ECE (Economic Commission for Europe) (1985), *Production and Use of Industrial Robots*, New York: United Nations, 131 pp., annexes.

ECE (1986), *Recent Trends in Flexible Manufacturing*, New York: United Nations, 314 pp.

Edquist, Charles and Jacobson, Steffan, in collaboration with Jethanandani, Kishore (1984), Automation in the engineering industries of India and the Republic of Korea against the background of the experience in some OECD countries, Research Policy Institute, University of Lund, mimeo, 40 pp.

English, Maurice (1983), *The European IT-Industry, A Short Overview*, Brussels: Commission of the European Communities, Information Technologies Task Force, 41 pp.

Ergas, Henry (1984a), *Why do Some Countries Innovate More than Others?* Brussels, CEPS Papers, no. 5, 71 pp.

Ergas, Henry (1984b), Corporate strategies in transition, in Alexis Jacquemin (1984), *European Industry: Public Policy and Corporate Strategy*, Oxford: Clarendon Press, pp. 327–43.

Ernst, Dieter (1983), *The Global Race in Micro Electronics, Innovation and Corporate Strategies in a Period of Crisis*, Frankfurt: Campus Verlag, 290 pp.

ETUI (European Trade Union Institute) (1982), *Negotiating Technological Change, Brussels, 100 pp.*

ETUI (1985), Flexibility and Jobs—Myths and Realities, Brussels, 157 pp.

European Federation of Pharmaceutical Industries' Associations (1984), *The European Pharmaceutical Industry and the Development of Biotechnology*, Brussels, 36 pp.

Evans, John (1984), The impact of microelectronics on employment in Western Europe, mimeo, 9 pp.

Farrands, Christopher (1983), *The Biotechnology Boom in Europe*, The Economist Intelligence Unit, European Trends, no. 3.

Feigenbaum, Edward, and McCorduck, Pamela (1984), *The Fifth Generation, Artificial Intelligence and Japan's Computer Challenge to the World*, New York: Signet paperback, 334 pp.

Fishlock, David (1982), *The Business of Biotechnology*, London: Financial Times Business Information Ltd, 156 pp.

Fitting, Franziska (1985), *Eine gigantische Hochzeit. Funktionen von EDS bei der Reorganisation des General-Motors-Konzerns*, in *Die Mittbestimmung (Monatszeitschrift der Hans-Böckler Stiftung)*, no. 7.

Franko, Lawrence G. (1983), *The Threat of Japanese Multinationals, How the West Can Respond*, Geneva: IRM/John Wiley, 148 pp.

Freeman, Christopher (1982), *The Economics of Industrial Innovation*, London: Frances Pinter, 250 pp.

Friebe, Klaus P., and Gerybadze, Alexander (eds.) (1984), *Microelectronics in Western Europe. The Medium Term Perspective 1983–1987*, Berlin: Erich Schmidt Verlag, 223 pp.

Friedrich, Werner, and Rönning, Gerd (1985), *Arbeitsmarktwirkungen moderner Technologien*, Köln: Institut für Sozialforschung und Gesellschaftspolitik.

Fröbel, Folker, Heinrichs, Jürgen, and Kreye, Otto (1979), *The New International Division of Labour*, New York: Cambridge University Press, Studies in Modern Capitalism, 406 pp., appendix.

Fröbel, Folker, Heinrichs, Jürgen and Kreye, Otto (1986), *Umbruch in der Weltwirtschaft. Die Globale Strategie: verbilligung der Arbeitskraft/Flexibilisierung der Arbeit/Neue Technologien*, Reinbek: Rowohlt Taschenbuch Verlag GmbH, 588 pp.

Gaag, P. van der (1985), *Het Co-makership, een organisatorische innovatie*, Rotterdam: Erasmus Universiteit, Research Memorandum Vakgroep Economische Organisatievormen, 117 pp., annex.

Gabrielli, Gilberto (1986), An overview of public–private alliances, Lake Como, Italy, paper presented at the conference on 'International Technology-based Alliances' Stanford Research Institute, April 21–3.

Ghertman, Michael (1982), *La Prise de Décision*, Paris, Presses Universitaires de France.

General Electric Company (1985), *GEC Research*, brochure, 30 pp.

Gillespie, Andrew *et al.* (1984), *The Effects of New Information Technology on the Less Favoured Regions in the Community*, Commission of the European Communities, Studies Collection, Regional Policy Series, no. 23, Brussels.

Gonzales-Virgil, Fernando (1985), New technologies, industrial restructuring and changing patterns of metal consumption, *Raw Materials Report*, **3** (3).

Gorbis, Marina and Yorke, Karen (1985), *Strategic Partnerships: A New Corporate Response*, SRI International, 91 pp.

Goto, Akiro (1982), Business groups in a market economy, *European Economic Review*, **19**, pp. 53–70, North Holland Publishing Company.

Groupe CGE (1985), *Memorandum commun sur la formation dans le Groupe CGE*, Paris.

Hack, Lothar, and Hack, Irmgard (1985), *Die Wirklichkeit, die Wissenschaft. Zum wechselseitigen Begründungsverhältnis von 'Verwissenschatlichung der Industrie' und 'Industrialisierung der Wissenschaft'*, Frankfurt: Campus Verlag, 687 pp.

Haklisch, Carmela (1986), *International Technical Agreements in the Semiconductor Industry*, New York: Center for Science and Technology Policy, New York University.

Halfmann, Jost (1984), *Die Entstehung der Mikroelektronik, zur Produktion technische Fortschritts*, Frankfurt: Campus Verlag, 222 pp.

Hills, Jill (1986), *Deregulating Telecoms. Competition and Control in the United States, Japan and Britain*, London: Frances Pinter, 220 pp.

Hingel, Anders J. (1982), *Social Change and Technology in Europe, Current Events in Scandinavia*, European Pool of Studies, Brussels, September, report no. 9.

Hirsch, Fred (1978), *The Social Limits to Growth*, Cambridge, Mass.: Harvard University Press.

Horn, Ernst-Jürgen (1982), *Management of Industrial Change in Germany*, Sussex European Papers, University of Sussex, No. 13.

Horn, Ernst-Jürgen, Klodt, Henning, and Saunders, Christopher (1985), Advanced machine tools: production, diffusion and trade, in Margaret Sharp (ed.) (1985a), pp. 46–86.

House of Lords (1984), *Session 1984–85*, 8th Report, Select Committee on the European Communities, Minutes of Evidence, London.

ICI Europe (no date), *The Innovators*, Everberg: Public Relations and Publicity Department, 52 pp.

Industriegewerkschaft Metall (1983), *'Maschinen wollen sie—uns Menschen nicht', Rationalisierung in der Metallwirtschaft*, Frankfurt, Kurzfassung, 93 pp.

Italtel (1985), *Innovazione technologica e opportunità occupazionali. Il caso Italtel*, Milan.

Jacquemin, Alexis (ed.) (1984), *European Industry: Public Policy and Corporate Strategy*, Oxford: Clarendon Press, 377 pp.

Jacquemin, Alexis, en collaboration avec Marleen Lammerant et Bernard Spinoit (1986), *Compétition européen et coopération entre entreprises en matière de recherche-developpement*, Brussel: Commission de Communautés Européennes, Université Catholique de Louvain, 112 pp.

Jasanoff, Sheila (1985), Technological innovation in a corporatist state: the case of biotechnology in the Federal Republic of Germany, *Research Policy*, **14**, pp. 23–38.

Jeandon, J. P., and Zarader, R. (1982), The economic and social impact of robotics in France, in EPOS, report no. 10, *Social Change and Technology in Europe: Robotics*, Brussels, pp. 101–46.

Jeandon, J. P., and Zarader, R. (1983), *Automation et emploi. Pour un vrai débat autour des vrais problèmes*, Futuribles, March.

Junne, Gerd, and Tulder, Rob van (1984a), *European Multinationals in the Robot Industry*, Amsterdam/Geneva, pilot study, 47 pp.

Junne, Gerd (1984b), *Der strukturpolitische Wettlauf zwischen den kapitalistischen Industrieländern*, Politische Vierteljahresschrift, June, 25. Jahrgang, heft 2, pp. 134–55.

Junne, Gerd (1987), Automation in the North: consequences for developing countries' exports, in James A. Caporaso (ed.), *A Changing International Division of Labor*, Boulder: Lynne Rienner: London: Frances Pinter, pp. 71–90.

Jürgens, Ulrich, Dohse, Knuth, and Malsch, Thomas (1986), New production concepts in West German car industry, in Stephen Tolliday and Jonathan Zeitlin (eds.), *Between Fordism and Flexibility, the International Automobile Industry and Its Workers (Past, Present and Future)*, London: Basil Blackwell.

Kaplinsky, Raphael (1982), *Computer Aided Design, Electronics, Competitive Advantage and Development*, London: Frances Pinter, UNIDO, 144 pp.

Kaplinsky, Raphael (1984), *Automation, the Technology and Society*, Essex: Longman, 197 pp.

Kaplinsky, Raphael (1985a), *Microelectronics and Employment Revisited*, report prepared for ILO world employment programme, 164 pp.

Kaplinsky, Raphael (1985b), Electronic-based automation technologies and the onset of systemofacture: implications for third world industrialization, *World Development*, March, pp 423–39.

Kern, Horst, and Schumann, Michael, (1984), *Das Ende der Arbeitsteilung? Rationalisierung in der industriellen Production*, München: C. H. Beck Verlag, 361 pp.

Kleinknecht, Alfred, (1984), Innovation patterns in crisis and prosperity: Schumpeter's long cycle reconsidered, Amsterdam, dissertation, 239 pp.

Kondratieff, N. D. (1926), Die Langen Wellen der Konjunktur, Tübingen, *Archiv für Sozialwissenschaft und Sozialpolitik*, **56**, pp. 573–ff.

Labarrère, Claude (1985), *L'Europe des Postes et des Télécommunications*, Paris: Masson, Collection Technique et Scientifique des Télécommunications, CNET, ENST, 256 pp.

Landes, David (1969), *The Unbound Prometheus*, Cambridge: Cambridge University Press.

Läpple, Dieter, Pennen, Ton van der, and Vlek, Ruud (1986), *De woning als 'werkplek'. Nieuwe en oude vormen van arbeid in en rond het huis*, Utrecht: Jan van Arkel, 238 pp.

Levie, Hugo, and Moore, Roy (coordination) (1984), *The Control of Frontiers: Workers and New Technology; Disclosure and Use of Company Information*, Oxford: Ruskin, Final Report and Summary Report including profiles of twenty case studies, 116 pp., annexes.

Lewis, Chris, and Kristiansen, Bjorn, (1985), Chemicals manufacture via biotechnology—the prospects for Western Europe, *Chemistry and Industry*, September, pp. 571–6.

Locksley, Gareth (1985), Information technology and capitalist development, *Capital and Class* no. 27, Winter, pp. 81–105.

Lorenz, Gert (1982), The diffusion of emerging technologies among industrial countries, in Herbert Giersch (ed.), *Emerging Technologies: Consequences for Economic Growth, Structural Change and Employment*, Kiel: Institut für Weltwirtschaft, pp. 171–88.

Louisse, Jim (1983), *Bewapening en werkgelegenheid*, OSACI/Pax Christi, Amsterdam.

Lüthje, Boy (1985), Telekommunikationsindustrie und Strukturwandel; Weltmarktkonkurrenz, industrielle Strukturveranderung und Staatsintervention im telecommunikationssektor in der BRD, MA thesis, Universität Frankfurt, Fachbereich Gesellschaftswissenschaften.

Mackenzie, Ian (1985), Advanced computing: the issues for Europe, London Business School, Centre for Business Strategy, mimeo.

Mackintosh, Ian (1986), *Sunrise Europe. The Dynamics of Information Technology*, Oxford: Basil Blackwell, 288 pp.

Mackintosh International (1983), *Assessment of Alternative Information Technology Strategies—IT Renaissance for Europe*, Luton, Final Report to the Commission of the European Communities, 265 pp.

Mackintosh International (1984), *Eurogrid—A Broadband Scenario Information Technology Renaissance for Europe*, Luton, Main Report and appendices, 398 pp.

Malerba, Franco (1985), *The Semiconductor Business, The Economics of Rapid Growth and Decline*, London: Frances Pinter.

Malsch, Thomas, Dohse, Knuth, and Jürgens, Ulrich (1984), 'Industrial robots in the automobile industry. A leap towards "Automated Fordism"?' Berlin: IIVG discussion paper, 64 pp.

Malsch, Thomas, Dohse, Knuth, and Jürgens, Ulrich (1985), Japanischer Vorsprung im Technologiewettlauf? Zum Robotereinsatz in der Automobil industrie, in Sung Jo Park (Hg), *Das japanische Management in der Praxis: Flexibilität oder Kontrolle im Prozess der Internationalisierung und Mikroelektronisierung*, Berlin: Express Verlag.

Malsot, J. (1980), Filières et effets de domination dans le système productif, *Annales des Mines*, Janvier.

Markusen, Ann R. (1984), High-tech jobs, markets and economic development prospects: evidence from California, in Peter Hall and Ann Markusen (eds.), *Silicon Landscapes*, London: Allen & Unwin, pp. 35–48.

McKinsey & Company (1983), *A Call to Action. The European Information Technology Industry*, Brussels, Commission of the European Community.

McLean, Mick and Rowland, Tom (1985), *The INMOS Saga*, London: Frances Pinter, 199 pp.

McMillan, Charles (1984), *The Japanese Industrial System*, Berlin/New York: Walter de Gruyter.

Mensch, G. (1979), *Stalemate in Technology. Innovations overcome Depression*, Cambridge, Ballinger.

Meyer, Arnoud de (1986), Large European manufacturers and the management of R & D, *R & D Management*, **16** (2).

Michalet, Charles-Albert (1985), *Les Multinationales face a la crise*, Geneva: IRM/PUF dossier, no. 4, 81 pp.

MITI (Ministry of International Trade and Industry) (1984), *Small Business in Japan*, Tokyo, Small and Medium Sized Enterprise Agency.

Moses, Vivian (1983), Microbes and oil recovery: an overview, in *Biotech 83. Proceedings of the International Conference on the Commercial Applications and Implications of Biotechnology*, Northwood Hills: Online Conference Ltd, pp. 415–22.

Murray, Fergus (1983), The decentralisation of production, the decline of the mass collective worker? *Capital and Class*, Spring, no. 19, pp. 74–99.

Nederlandse Middenstandsbank NV (1985), *De relatie tussen grote en kleine bedrijven in de industrie*, Amsterdam, March.

Noble, David (1984), *Forces of Production*, New York: Alfred A. Knopf, 409 pp.

OECD (Organisation for Economic Corporation and Development) (1979), *The Impact of the Newly Industrializing Countries on Production and Trade in Manufactures*, Report by the Secretary General, Paris.

OECD (1981), *Microelectronics, Productivity and Employment*, Information, Computer, Communication (ICCP), no. 5, Paris, 292 pp.

OECD (1982), *Microelectronics Robotics and Jobs*, ICCP, no. 7, Paris, 265 pp.

OECD (1983a), *Telecommunications, Pressures and Policies for Change*, Paris.

OECD (1983b), *Industrial Robots. Their Role in Manufacturing Industry*, Paris, 93 pp.

OECD (1983c), *Positive Adjustment Policies. Managing Structural Change*, Paris.

OECD (1983d), *Transborder Data Flows: An Overview of Issues*, DSTI/ICCP/83.29, Paris.

OECD (1984), *Trade in High Technology Products: An Examination of Trade Related Issues in the Machine Tool Industry*, Paris, DSTI/SPR 8340, 53 pp.

OECD (1985a), *Economic Outlook*, nr 37, Paris.

OECD (1985b), *Structural Adjustment and Multinational Enterprises Programme*, Paris, 68 pp.
OECD (1985c), *The Semiconductor Industry. Trade Related Issues*, Paris, 145 pp.
OECD (1985d), *Software: An Emerging Industry*, ICCP9, Paris.
OECD (1986a), *Technical Co-operation Agreements between Firms: Some Initial Data and Analysis*, DSTI/SPR/86.20, Paris, part 1, part II, 87 pp.
OECD (1986b), *Science and Technology Indicators*, no. 2, Paris.
OECD (1986c), *Review of National Science and Technology Policies: The Netherlands*, Paris.
OECD (1986d), *Review of Innovation Policies: France*, Paris, Directorate for Science, Technology and Industry.
Ohmae, Kenichi (1985), *Triad Power. The Coming Shape of Global Competition*, London and Basingstoke: Collier Macmillan.
OTA (Office of Technology Assessment) (1983), *International Competitiveness in Electronics*, Washington DC: Congress of the United States, 546 pp.
OTA (1984a), *Commercial Biotechnology. An International Analysis*, Washington DC: Congress of the United States.
OTA (1984b), *Computerized Manufacturing Automation: Employment, Education, and the Workplace*, Washington DC: Congress of the United States, OTA–CIT 235, 471 pp.
OTA (1985), *Information Technology R & D, Critical Trends and Issues*, Washington DC: Congress of the United States, 342 pp.
Pannenborg, A. E. (1986), Internationale Samenwerking: ESPRIT, in *Informatie en Informatiebeleid*, vierde jaargang, no. 2, pp. 23–7.
Patel, Pari, and Pavitt, Keith (1986), *Measuring Europe's performance: results and prospects*, Henry Ergas (ed.), *A European Future in High Technology*, Brussels: Centre for European Policy Studies.
Patel, Pari, and Pavitt, Keith (1987), Is Europe losing the technology race? *Research Policy*, no. 1.
Pelissolo, Jean-Claude (1980), La Biotechnologie, demain? Rapport à Monsieur le Premier Ministre, Paris: Documentation Française.
Petre, Peter (1985), The hightech car hits the road, *Fortune*, 29 April, pp. 114–21.
Philips Gloeilampenfabrieken, NV (1985), *An Agenda for Action — 1990*, Report to the Commission on the European Communities, Brussels.
Piore, Michael J., and Sabel Charles, F. (1984), *The Second Industrial Divide. Possibilities for Prosperity*, New York: Basic Books, 355 pp.
Piper, Allan (1985), Learning to compete: two telephone monopolies take the plunge into the real world, *International Management*, April, pp. 32–40.
Porat, M. U. (1977), *Information Economy*, Washington DC. Volume I: *Definition and Measurement*.
President's Commission on Industrial Competitiveness (1985), *Global Competition, the New Reality*, Washington DC: Government Printing Office, Volumes I and II.
Reich, Robert B. (1983), *The Next American Frontier, a Provocative Program for Economic Revival*, New York: Penguin, 324 pp.
Rhodes, Martin (1985), *French Government Subsidies for Information Technology*, paper presented at the ECPR joint sessions of workshops, Barcelona, April, 30 pp.
Richardson, Jeremy J. (1986), Policy, politics and the communications revolution in Sweden, *West European Politics*, October.

Richardson, Jeremy J., Moon, Jeremy, and Webber, Douglas (1984), Information technology, employment and economic recovery: the role of the British, French and West German governments, paper prepared for RIPA conference 'The Shifting State: Rules, Roles and Boundaries in the 1980s', University of Aston, 28 pp.

Richman, Louis S. (1986), Daimler-Benz Conglomerates, *Fortune*, 15 September, pp. 48–52.

Rodenburg, N. (1980), *Herstrukturering en Philips*, Economisch Statistische Berichten (Rotterdam), March.

Roobeek, Annemieke J. M. (1986a), Biotechnology and regulation: different perspectives in the United States and the European Community, *Trends in Biotechnology*, **4**, (4), April, pp. 75–8.

Roobeek, Annemieke J. M. (1986b), Internationale Aspecten bij de opkomst van nieuwe materialen, *Internationale Spectator*, December, jrg. **40–12**, pp. 752–60.

Roobeek, Annemieke J. M. (1987a), *De rol van de technologie in de economische theorievorming*, Amsterdam: Scheltema Holkema Vermeulen.

Roobeek, Annemieke J. M. (1987b), The crisis in Fordism and the rise of a new technological paradigm, *Futures*, April, pp. 129–54.

Rothwell, Roy (1985), Technology policy in Britain, paper for conference on 'A Competitive Future for Europe?', Rotterdam, December.

Rothwell, Roy, and Zegveld, Walter (1980), *Innovation and Technology Policy*, London: Frances Pinter.

Rothwell, Roy, and Zegveld, Walter (1982), *Innovation and the Small and Medium Sized Firm*, London: Frances Pinter, 268 pp.

Rothwell, Roy, and Zegveld, Walter (1985), *Reindustrialisation and Technology*, London: Longman.

Ruivenkamp, Guido (1986), De invoering van biotechnologie in de agroindustriele produktieketen: een analyse van Shell's positie, Amsterdam, manuscript for book on Shell by Stichting Onderzoek Multinationale Ondernemingen (SOMO).

Sable, Charles F. (1982), *Work and Politics. The Division of Labour in Industry*, Cambridge: Cambridge University Press.

Salisch, Heinke (rapporteur) (1985), *The Impact of Microtechnology on Job Opportunities for Women*, European Parliament, working documents, 1985–1986, PE 100.121/fin, 27 pp.

Sargeant, Ken (1984), Biotechnology, connectedness and dematerialisation: the strategic challenges to Europe, and the community response, Paper presented at 'Biotechnology '84'', organized by the Royal Irish Academy and the Society for General Microbiology at Dublin.

Sasson, Albert (1983), *Les Biotechnologies, défis et promesses*, Paris: Unesco, 336 pp.

Savary, Julien (1984), *French Multinationals*, Geneva: IRM.

Scharpf, Fritz W. (1985), Strukturen der post-industriellen Gesellschaft, oder: Verschwindet die Massenarbeitslosigkeit in der Dienst, Leistungs-, und Informations-Okonomie?, Berlin: IIMV discussion paper.

Schmitz, Hubert (1985), Microelectronics: implications for employment, outwork, skills and wages, Institute for Development Studies, Sussex University, Discussion Paper, 42 pp.

Schumpeter, Joseph A. (1939), *Business Cycles: A Theoretical, Historical and Statistical Analysis of the Capitalist Process*, New York: McGraw-Hill, 2 volumes.

Schutte, Giorgio, *et al.* (1986), EUREKA, determinanten en consequenties van Europees Technologie Beleid, Universiteit van Amsterdam, mimeo, June.

Scott, Roger (1984), Design and evaluation of experiments to provide scale-up information for the ICI single cell protein process, in *Biotech 83. Proceedings of the International Conference on the Commercial Applications and Implications of Biotechnology*, Northwood Hills: Online Conferences Ltd.

Semiconductor Industry Association (1983), *The Effect of Government Targeting on World Semiconductor Competition. A Case History of Japanese Industrial Strategy and its Costs for America*, San José/Washington, 109 pp., annexes.

Servan-Schreiber, Jean-Jacques (1967), *Le Défi américain*, Paris: Denoël.

Shaiken, Harley (1985), *Work Transformed. Automation and Labor in the Computer Age*, New York: Holt, Rinehart & Winston, 306 pp.

Sharp, Margaret (ed.) (1985a), *Europe and the New Technologies*, London: Frances Pinter.

Sharp, Margaret (1985b), *The New Biotechnology: European Governments in Search of a Strategy*, University of Sussex: European Paper, no. 15.

Sharp, Margaret, and Claire, Shearman, (1987), *European Technological Collaboration*, London: Chatham House Papers, no. 36, Royal Institute of International Affairs, 122 pp.

Shell (undated), *Science & Technology*, London, different issues.

Siemens AG (1984), *Bildungsarbeit im Unternehmen*, Fürth, 24 pp.

Sitter, L. U. de (1982), *Op weg naar nieuwe fabrieken en kantoren*, Deventer: Kluwer, 258 pp.

Sobel, Robert (1983), *IBM: Colossus in Transition*, London, Bantam Books.

Steinbrinck, Renate (1985), *Elektronischer Heimarbeit, Neue Informations- und Kommunkationstechnologien und die Zukunft der Arbeit: Dezentralisierung ins telematisierte Zuhause*, Brussels, European Parliament, 91 pp.

Stopford, John, and Dunning, John (1983), *Multinationals—Company Performance and Global Trends*, London: Macmillan.

Stopford, John, and Turner, Louis (1985), *Britain and the Multinationals*, Geneva: Wiley/IRM series on multinationals, 282 pp.

Swan, Denis (1983), *Competition and Industrial Policy in the European Community*, London, Methuen's EEC series.

Thorn EMI Central Research Laboratories (undated), *Research Achievements by Thorn EMI Central Research Laboratories*, Middlesex.

Truel, J. L. (1980), *L'Industrie mondiale des semi-conducteurs*, Thèse de Troisième Cycle, Paris, IX.

Tsurumi, Yoshi (1984), The challenges of the Pacific age, *World Policy Journal*, Fall, pp. 63–87.

Tulder, Rob van (1987), *The 'Factory of the Future', Productivity and Income, a Scenario for Metalworkers*, Geneva: International Metalworkers' Federation, 61 pp.

Tulder, Rob van, and Empel, Eric van (1984), *European Multinationals in the Semiconductor Industry. Their position in Microprocessors* Amsterdam/Geneva, pilot study for the Institute for Research and Information on Multinationals (IRM) 81 pp.

UNCTC (United Nations Centre of Transnational Corporations) (1984a), Transnational corporations in biotechnology, rev. draft, 11/84, New York.

UNCTC (1984b), *The Role of Transnational Corporations in Transborder Data Flows*, New York, E/C.10/1984/14.

United Nations (1980), *1980 Yearbook of World Energy Statistics*, New York.

Uri, Pierre, (rapporteur) (1971), *Rapport sur la capacité concurrentielle de la Communauté Européenne*, Brussels: Commission of the European Communities, 85 pp.

US Department of Commerce, International Trade Administration (1983a), *An Assessment of U.S. Competitiveness in High Technology Industries*, Washington DC, 68 pp.

US Department of Commerce (1983b), *High Technology Industries: Profiles and Outlooks: The Robotics Industry*, Washington DC, 54 pp. (Also in the series: the computer industry, telecommunications industry, semiconductor industry.)

US Department of Commerce (1984a), *A Competitive Assessment of the U.S. Manufacturing Automation Equipment Industries*, Washington DC, prepared by the Industry Analysis Division, 101 pp.

US Department of Commerce (1984b), *A Competitive Assessment of the U.S. Software Industry*, Washington DC, 77 pp., annexes.

US Department of Commerce (1985), *A Competitive Assessment of the US Semiconductor Manufacturing Equipment Industry*, Washington.

Webber, Douglas, Moon, Jeremy, and Richardson, J. J. (1984), *State Promotion of Information Technology in France, Britain and West Germany*, Glasgow: Strathclyde Papers on Government and Politics, no. 33.

Webber, Douglas (1986), Die ausbleibende Wende bei der Deutschen Bundespost. Zur Regulierung des Telekommunikationswesens in der Bundesrepublik Deutschland. *Politische Vierteljahresschrift*, **27**, no. 4 (December), p. 397–414.

Welzk, Stefan, (1986), *Boom ohne Arbeitsplätze*, Köln: Kiepenheuer & Witsch, 251 pp.

Williamson, Oliver E. (1975), *Markets and Hierarchies: Analysis and Antitrust Implications, A Study in the Economics of Internal Organization*, New York/London: The Free Press, 286 pp.

Wiltgen, Paul (1987), Research and development for high technology in European multinationals, University of Amsterdam, Department of International Relations, mimeo, 140 pp.

Wit, Andries de (1985), Flexibilisering en nieuwe technologie, Eindhoven, mimeo, 108 pp.

Yankee Group (1983), *The Report on European Telecommunications*, Watford, Volume Two.

Yoxen, Edward J. (1985), Government promotion of biotechnology, *Physical Technology*, **16**, pp. 234–42.

Index

Aakvaag, T. (Norsk Hydro) 214
Acceptance of technologies 146, 147, 206
Accumulation x, 2, 19, 26, 29, 259
ACMA (Renault) 47, 87
ACME (Applications of Computers to Manufacturing Engineering) 170
Acorn 43
Acquisitions 33, 35–37, 39, 43, 46, 52, 56, 59–62
Administration xi, 11, 21, 110, 111
Advanced Micro Devices (AMD) 35
AEG (Allgemeine Elektricitäts–Gesellschaft) 29, 30, 32, 34, 45, 64, 66, 68, 72, 95, 160, 178, 182, 187, 193, 200, 214, 228, 230–233
Aerospatiale 51, 52
AFRI (Association Française de Robotique Industrielle) 77, 109, 138, 260
Agfa-Gevaert 64
Aglietta, M. 29, 260
Agnelli, U. (Fiat) 214
Agricultural and Food Research Council 174
Agrochemical companies 59, 60, 143
Aircraft industry 6, 7, 14, 29, 62
AKZO 52, 62, 64, 69, 72, 95, 143, 177, 178, 217, 228, 241
Alcatel-Thomson 44
Alfa-Laval 61, 179
Allen-Bradley 49, 80, 172, 238
Alliances xiii, xiv, 221, 232
Allied-Bendix 72
Alloys 30
Altshuler, A. 31, 81, 260
Alsthom-Atlantique 49, 193
Alvey project 166, 167, 224
Amada 237
AMC (American Motors Corp.) 78
AMD 34, 238, 239
AMES (Automatisation et Mutations Economiques et Sociales) 78, 109, 260

AMI 34
Amperex 35
Amylum 177
Anastassopoulos, J.-P. 260
Anderson, M. 260
Andries, M. 260
ANT 235
Antibiotics 12, 57
Anti-Trust 41, 164, 198–201, 254
Anova 214
Apple 11
Applicon 51
Applied Biosystems 142
APT 190
ARA Programme (France) 170
Aramide fibre 62
Arnold, E. 10, 50–52, 260
Artificial intelligence 16, 53, 114
Asahin Glass 229, 241
Asbestos 62
ASEA 49–51, 64, 66, 69, 95, 104, 120, 140, 172, 179, 193, 214, 216, 228, 237, 238, 245
Ashai Chem. 240
Assembling 20, 48, 74, 76, 77, 79–81, 96, 97, 102, 105, 107, 110, 113, 115, 119, 120, 140
Associations Européenne de Robotique 138
ATEA-Siemens 190
ATES 33
Atlantic Richfield 71, 72
AT&T (American Telephone and Telegraph Corp.) 11, 16, 43–46, 54, 66, 90, 135, 178, 190, 201, 203–205, 216, 227–229, 235, 236, 239, 243, 252
Austerity policy 184
Austin Rover 81
Australia 168
Austria 162, 168, 190, 229
Automation ix, 3, 9, 15, 17, 19, 23, 31, 47, 51, 53–55, 63, 75–77, 80, 82, 83, 86, 87, 94, 96–98, 102, 103,

271

Automation *(continued)*
　105–111, 116, 117, 119, 120, 125,
　138, 141, 165, 169–172, 222, 249
　flexible 9, 23, 25, 26, 47, 138
　rigid 3, 4, 22
Automobile industry 1, 22, 29–31, 47,
　48, 50–52, 61, 63, 67, 76–82,
　108–110, 113, 114, 139–141, 230
Autotrol 51
Auxilec 49

BASF AG 41, 42, 58, 64, 67, 68, 95,
　96, 120, 173, 174, 228, 242, 243
Balance of payments xiii
Banking 103
Banque de Suez 205
Banville, E. de 79, 261
Barnevik, P. (ASEA) 104
Basic research 143, 144, 147, 148
Basic technologies 1, 2, 6
Batch size 117
BAT Industries 70, 214
Battlefield-of-the-future 17
Bauer, M. 261
Baxendeel, P. 215
Bayer AG 57, 64, 66–68, 71, 95, 113,
　143, 173, 174, 178, 228, 260
Beffa, J.-L. (Saint-Gobain) 214
Belgian Institute for Regulation and
　Automation 139
Belgium 39, 44, 78, 119, 136, 162, 168,
　170, 173, 186, 188, 190, 229, 231
Bell Laboratories 90
Bell Telephone Company (Belgium)
　44, 72, 168
Bendix Corp. 47, 238
Berendsen, H. J. C. 142, 260
Bessant, J. 260
Bethlehem Steel 127
Bianchi, P. 81
Biochips 16, 19
Biogen 241
Bio-informatics 16, 71, 142
Biomass 59, 60
Biomolecular Engineering Programme
　(BEP) 175, 176
Bioprocess-technology 13–16
Bioreactors 58
Biotechnology ix, 6, 11–16, 22, 24–26,
　50, 56–61, 68, 69, 71, 72, 83, 89,
　114, 141–146, 172–177, 181, 185,
　207, 222, 241, 242, 244, 248

Biotechnology Regulation Interservice
　Committee (BRIC) 145
Biotronics 219
Blanc, G. 260
BMFT, *see* Bundesministerium für For-
　schung und Technologie
BMW (Bayerische Motoren Werke AG)
　29, 47, 65, 67, 78, 95, 140
Boeing Aerospace 236
Boeringer Ingelheim 242
Bolwijn, P. T. 149, 260
Bosch GmbH 35, 45, 48, 64, 68, 72,
　79, 81, 82, 91, 95, 194, 203, 214,
　216, 228, 230, 235–237
Boucher, E. le 205
Bourns 34
Boyer, R. 99, 260
Breitschwerdt, W. (Daimler–Benz) 29,
　214
Breweries 60, 144
Brinkman, S. 149, 260
Brite 222–233
British Aerospace 39, 52, 235, 236
British Leyland 31, 70, 78, 232, 240
British Petroleum (BP) 55, 59, 64, 67,
　68, 71, 72, 95, 96, 174, 179, 228
British Robot Association 138
British Telecom (BT) 38, 39, 44, 65, 67,
　69, 95, 167, 189, 192, 195, 199,
　203–205, 228, 232, 235, 240
Broadband communication 9
Brown, Boveri & Cie. (BBC) 39, 49,
　72, 179, 195
BSN–Gervais 215
Bull, A. T. 12, 260
Bull (Machine Bull) 39–42, 54, 55, 65,
　67, 69, 85, 113, 127, 167, 181, 193,
　196, 203, 213, 216, 228, 230–233,
　236, 239, 242, 243, 246
Bundesministerium der Finanzen 162,
　260
Bundesministerium für Forschung und
　Technologie (BMFT) 16, 98, 111,
　164, 165, 171, 174, 176, 182, 183,
　193, 200, 260
Bundesministerium für Wirtschaft 103,
　200, 260
Bundespost 136, 189, 205
Bureaucratization 84, 146
Burroughs 40, 41
Business associations 173
Business cycle ix, 8

Buttel, F. H. 260
Bylinsky, G. 260

Cable & Wireless 72, 204
CAD/CAM (Computer-Aided Design/
  Computer-Aided Manufacturing)
  8, 10, 14, 16, 26, 50–53, 71, 80, 81,
  85, 87, 102, 109–111, 119, 138, 141,
  142, 170, 172, 225, 226, 230, 246,
  251
CADIS System 51, 52
Calgene 241
Calma 51, 52
Cambridge Interactive Systems 52
CAP Gemini Sogeti 55
Capital goods 5, 21
Capital intensity 2, 3, 94–96
Caporaso, J. A. 264
Captive production 31
Cardo 61
Carlsson, J. 168, 261
Carriers of technology ix, xiii
Cartelization 199, 234
Cash registers 20
Cawson, A. 189, 197, 261
CEA (Atomic Energy Commission,
  France) 160
Cell fusion 13, 14, 68
Celltech 174, 177, 241
CEM 49
Cement 62
Central Institute for Industrial Research
  (Norway) 52
Centralization xi, 74, 75, 85–87,
  90–92, 117–119, 152
Central Processing Unit (CPU) 9–11
Centre de Prospective et d'Évolution
  (CPE) 80, 81, 262
Ceramics 15, 24, 30, 31, 61, 62, 71
Ceramtech 62
Cetus 241
CGCT 190, 201, 203
CGE 42, 44–46, 49, 65–68, 72, 112,
  113, 135, 160, 168, 169, 172, 178,
  181, 182, 193, 195, 196, 205, 213,
  216, 226, 228, 230–233, 245
CGMS 49
Chanaron, J. J. 79, 261
Chemical industry 14, 56–58, 60–63,
  67, 71, 90, 120, 126, 143, 144, 207
Chenard, A. 103, 261
Chesnais, F. 247, 261

Chevènement 165
China 168
Chips, *see* Semiconductors
Chirac 165, 201, 205
Chrysler 78, 140
Ciba–Geigy 58, 64, 67, 68, 95, 127,
  143, 179, 216, 228, 240, 241
Ciborra, C. 43, 261
CII–Honeywell–Bull 42, 160
CIM (Computer Integrated Manu-
  facturing), *see* CAD/CAM
CIT–Alcatel 35, 46, 168, 181, 190, 200,
  203, 213, 236, 237
Civil servants 184, 185, 207
Claire, S. 269
Clark, J. (Plessey) 214
Clothing 2
Clusters of technology 6, 11, 13–18,
  30, 32, 33, 71, 249, 253
CNET (Centre National d'Etudes des
  Télécommunications) 189
Cohen, E. 261
Cohendet, P. 58, 261
Comakership 93, 94
Comau (Fiat) 47, 49, 140
Commercialization ix, 13, 43, 129, 142,
  144, 147, 149, 155, 157, 199, 200
Commission of the European
  Communities 103, 139, 140, 143,
  149, 173–176, 200, 213, 234, 259,
  261
Commodity prices 4
Common Agricultural Policy (CAP)
  144, 145
Communication xi, 3, 21, 22, 26
Commuting 4, 21
Compagnie Française des Pétroles 70
Company culture 33, 71
Compatibility 90, 219
Competition ix, xi–xiii, 2, 45, 52, 61,
  156, 158, 198, 234, 254, 257
Competitiveness xii, 3, 66, 106, 118,
  122, 125–155, 164, 169, 187, 201,
  202, 207, 209, 220, 255
Composites 15, 25, 62, 63, 185
Computers 6–10, 16, 21, 32, 40–44, 68,
  69, 84, 85, 100, 105, 159, 242, 243
Computervision 51, 52
Concentration 32, 36, 43–45, 47, 51,
  54, 55, 73, 80, 81, 108, 117, 139,
  177–185, 198, 200, 201, 254
Concord Data Systems 235

Conflicts xiii, xiv, 254
Constantia Industrieverwaltung GmbH 214
Consumer electronics 2, 5, 8, 14, 131, 134
Consumer demand 5
Consumption pattern ix, 5, 29, 131
Control xi, 21, 74, 75, 81, 85, 89, 90, 117, 153, 196, 207, 259
Control Data 235, 238
Convergence of technologies xi, xii, 15, 18, 19, 82, 89, 90
Coombs, R. W. 74, 75, 261
Cooperation xiii, xiv, 33, 35–39, 42–50, 56, 63, 70, 82, 93, 128, 130, 148, 155, 172, 173, 177, 184, 185, 199, 200, 208, 209–253, 257, 258
Coordination 83, 84, 91
COPICS software 54
Core technologies x, xii, 6, 7, 11, 15, 17, 19, 28, 30, 63, 66, 67, 70–73, 82, 87, 121, 125, 128, 163, 166, 168, 169, 172, 185, 253
Coriat, B. 262
Corning Glass 218, 240
COST 212, 223
Cost of labour 2, 5
Cost-plus basis 33, 186
Court des Comptes 182
Cowan, J. T. 260
CPE (Centre de Prospective et d'Évolution) 80, 81, 262
Creusot Loire 182
Crisis ix, 1, 5, 23, 28, 29, 156–158, 162
Critical mass 17, 89, 257
Cross-subsidization 205
CSM 60
Curien 165
Currency fluctuations 3
Customized chips 39–40

DAF 87, 93
Daihatsu 79
Daimler–Benz 29–31, 47, 51, 65, 67, 68, 78, 79, 95, 140, 178, 187, 193, 200, 214, 216, 228, 230–233, 254
Dainichi Kiko 49, 237, 238
Dairy industry 144
Dalgetty 70
Dang Nguyen, G. 190, 204, 205, 262
Dankbaar, B. 165, 262

Danske Sukkerfabrikker 60
Datel 160
Dassault 51, 182
Davidson, W. H. 11, 262
Davignon, E. 214, 215, 222
Data processing industry, see Computers
Datasaab 72, 83, 168
Datavision 172
Datbit 35
De Benedetti, C. (Olivetti) 42, 214, 220
Decentralization xi, 3, 21, 24, 25, 75, 88–91, 117, 119
DECHEMA (German Society for Chemical Engineering) 173, 177
Decision making xii, 85, 86, 90, 91, 97, 110, 196
Defence, see Military
De Forenede Bryggerier 214
Degussa 64, 69, 95, 96, 176, 228, 240
Dekker, W. (Philips) 12, 214
Delapierre, M. 262
Delors, J. 215
Demand pull 40, 131, 134
Dematerialization of production 23
Demonstration plants 104, 170
Denden family 203
Denmark 56, 60, 119, 136, 162, 170, 173, 186, 190, 229, 231
Depreciation 21
Deregulation 45, 136, 197–206, 257
Design xi, xii, 9–11, 24, 25, 39, 50, 51, 74, 79, 81, 89, 91, 97, 100, 102, 106, 110, 154, 155
Devagnet 165
Developing countries, see Third World
Development time 26
DGT (Direction Generale de Télécommunication) 189, 205
Dickson 35
Dieli 167
Diffusion xii, 10, 75, 77, 78, 80, 100, 108, 109, 111, 119, 122, 128, 130, 140, 151, 153, 154, 163, 166, 168, 170, 171, 257
Digital Electronic Automation (DEA) 49
Digital Equipment Co. 11, 40, 41, 80, 91, 94, 127, 227, 238, 243
Digitalization 8, 11, 16, 19, 134, 135, 188

Diversification x, xii, 5, 29–31, 35, 39, 42, 45, 48, 50, 61, 62, 66, 67, 71, 72, 74, 81, 82, 126, 128, 221, 253
Division of labour 73, 257
Divisions xii, 75, 82–84, 86, 154
DNA code 12, 13
Dohse, K. 76–78, 109, 110, 113, 264, 266
Dornier 29, 30, 187, 200
Dosi, G. 132, 159, 161, 262
Dow Chemical 127
Downstream technologies 13, 15
Downtime 22
Doz, Y. 234, 246, 262
Drugs 6
DSM 62, 64, 67, 68, 95, 114, 115, 175, 197, 228, 240, 241, 245
DSR 49
Dunning, J. 195, 269
Dupont 62, 127, 235, 240
Durham K. 215
Dussauge, P. 260

ECE (European Commission for Europe) 139, 234, 262
Economies of scale 3, 26, 47, 50, 60, 73, 81, 82, 89, 202, 211
ECRC (European Computer Industry Research Centre) 85
Edible oils 60
Edquist, C. 110, 262
EDS (Electronic Data Systems) 29, 55, 72, 80, 85, 86, 91, 187, 259
Education: *see* Qualification, Training Efficiency
Electrolux 50, 72
Electronic Arrays 34
Electronics industry 7, 12, 32, 47, 61, 66, 71, 77, 81, 105, 120, 126, 141, 159–162, 165, 182, 187, 190, 191, 196, 207, 230
Elf Aquitaine 59, 64, 67, 68, 71, 90, 95, 96, 143, 175, 177, 178, 228
Eli Lilly 58, 127
Ellemtel Development Co. 191
Empel, E. v. 234, 269
Employment xii, xiii, 5, 10, 19–21, 79, 97–112, 120, 122, 123, 125, 203, 207, 254, 255, 259
Energy consumption 2, 4, 8, 19, 23–25
English, M. 262
ENI 70, 179

Engineering 56, 61, 173
Enka 240
Entrepreneurial spirit 43, 147
Environment 118
Enzymes 12–14, 56, 57, 143, 145, 174
Ergas, H. 130, 148, 151
Ericsson 37, 41, 42, 44–46, 51, 53, 54, 65–67, 69, 72, 83–85, 91, 95, 104, 135, 136, 168, 179, 183, 190, 191, 193, 195, 201, 203, 228, 235-237, 239, 243, 248
Ernst, D. 262
Esprit 196, 213–215, 222–233, 246, 251
Ethanol 58
ETUI (European Trade Union Institute) 113, 262
Euclyde System 172
Euratom 211
EUREKA 39, 200, 212, 223–233, 244, 247, 259
European programmes xiii, 222–231
European Community xiii, 1, 211–216
European Federation of Biotechnology 177
European Federation of Pharmaceutical Industries' Associations 56, 262
European Silicon Structures (ES2) 39
Eurotechnique 160
Euroventures 215, 216, 230
Evans, J. 102, 105, 262
Evans & Sutherland (US) 52
Export controls 37
Exports 77, 133, 134
Exxon 34, 71, 72, 127

Fabius 165
Facit 42, 72, 83
Facsimile 45
Factory-of-the-future x, 17–19, 66, 83, 88, 102, 219
Factory size 3, 26
Factory space 22, 25
Fairchild 34, 35, 132, 236
Faltas, S. 187
Fanuc 29, 48
Farm-of-the-future x, 17–19, 102
Farrands, C. 174, 262
Feigenbaum, E. 10, 262
Fermentation technology 12–16, 58, 60, 61, 68, 143–145
Ferranti 32, 34, 35, 45, 50, 161, 167, 187, 239

Feruzzi 58, 214
Fiat 31, 46, 47, 49, 50, 58, 65, 67, 68, 77–81, 87, 95, 140, 179, 193, 216, 228, 240
Fifth generation computers 10, 166, 167, 184
Filière electronique 166
Filière productique 171, 172
Finland 191, 229
Fishlock, D. 263
Fitting, F. 86, 263
Flavours 60
Flexibility 4, 5, 9, 19, 75, 77, 84, 94, 107, 115,123, 153, 200, 253
Fluor Elastomers 62
Fokker 52
Followers Strategy 127–129
Fonds industriél de Modernisation (FIM) 170, 172
Food processing 14, 56, 60, 67, 144, 177, 230
Ford Motor Corp. 31, 77,78, 81, 91, 122, 140, 178, 229
Fordism 29
Foster, G. 168
Foxboro 229
Fragrances 60
France 16, 32, 33, 38, 40, 41, 44, 46, 47, 49, 51, 52, 54, 55, 59, 60, 77, 78, 80, 103, 105, 112, 114, 119, 122, 130, 131, 136, 138, 139, 142–144, 147, 151–153, 158–160, 162–167, 170–176, 178, 180–182, 185–189, 190, 196, 200, 201, 203, 205, 207, 229, 231, 233, 247, 256
Franko, L. G. 8, 263
Fraunhofer Gesellschaft 171
Freeman, C. 6, 159, 263
Friebe, K. P. 263
Friedman, M. 1
Friedrich, W. 98, 263
Fröbel, F. 263
Fuji 239
Fujitsu 34, 41, 135, 240, 242, 243, 246
Fujitsu Fanuc 48, 237
Furnas Electricas de Cataluna 214

Gaag, P. 93, 94, 263
Gabrielli, G. 234, 244, 263
GALIA (Groupement pour l'Amélioration des Liaisons dans l'Industrie Automobile) 80

Gallium arsenide 15, 61, 185
Gambro group 61
Gardini, R. (Feruzzi) 214
Gel technology 62
Gene machines 16
Genentech 241, 242
General Agreement on Tariffs and Trade (GATT) 3
General Electric Company (GEC) 32, 35, 42, 44, 46, 48–50, 65–67, 161, 167, 179, 187, 189, 190, 193, 203, 213, 214, 216, 228, 230, 235, 237, 239, 240, 263
General Electric Inc. (GE) 16, 17, 34, 51, 52, 127, 178, 205, 236, 237, 245
General Motors (GM) 29, 77, 78, 80, 85, 86, 127, 140, 178, 180, 187
General Motors Fanuc 29
Genetic engineering 12, 58, 142–145, 173
Genex 242
Gerber 51
Germany, *see* West Germany
Gerybadze, A. 263
Ghertman, M. 57, 263
GI 34
Gillespie, A. 91, 119, 263
Giscard D'Estaing 171
Gist–Brocades 56, 64, 69, 70, 95, 143, 175, 177, 228, 241
Glaxo 143, 174
Gomez, A. (Thomson) 38, 214
Gonzales-Virgil, F. 31, 263
Gorbis, M. 234, 263
Goto, A. 92, 263
Gould 34
Government ix, xii, xiii, 35, 37–39, 42, 44, 47, 50, 62, 70, 82, 109, 111, 131, 136, 137, 142, 144, 148, 149, 155–211, 227
Grado 52
Grand Corps 197
Grandes écoles 185
Grand Metropolitan 70
Greece 173, 176, 191, 229, 231
Growth ix, xiii, 1, 2, 4, 6, 26, 27, 97, 101, 106, 108, 123, 124, 162, 187
Grundig 160, 200
GTE 46, 135, 190, 203, 229, 235, 236, 239
Guy, K. 260
Gyllenhammer, P. (Volvo) 214, 215

Hack, I. 263
Hack, L. 263
Hagedorn, J. 183
Haklisch, C. 234, 263
Halfman, J. 7, 263
Hall Automation 49
Hamel, G. 234, 246, 262
Hamilton, G. viii
Harris 34, 39, 239
Hart, J. viii, 84
Harvey-Jones, J. 215
Hasler 168
Hattori group 72
Hayes, P. 214
Headquarters xi
Heineken 60
Heinrichs, J. 263
Hen, P. E. de 200
Hercules 241
Hewlett-Packard 94, 243
Hierarchization 76, 79, 80, 92, 94
Hifi systems 105
High chem 62
Hills, J. 204, 264
Hingel, A. J. 168, 264
Hirsch, F. 5, 264
Hitachi 34, 41, 42, 49, 66, 82, 127, 132, 178, 237, 242, 246
Hoechst AG 57, 62, 64, 67, 68, 95, 127, 143, 173–175, 177, 178, 228, 240
Hoffmann-LaRoche 58, 95, 143, 179, 241
Holland, *see* Netherlands
Holmes, P. 189, 197, 261
Holt, G. 260
Home labour 79, 94, 114, 115, 121
Honda 79, 81
Honeywell 41, 42, 229, 236, 246
Hoogovens 63, 87, 238
Horn, E.-J. 171, 182, 183, 264
Households 17, 67
House of Lords 264
Hughes Aircraft 29, 34, 187, 236
Human growth hormones 61
Human resources 150–155

IBM 11, 16, 40, 41, 43–45, 49–52, 54, 55, 66, 80, 84, 87, 90, 91, 93, 94, 108, 127, 159, 178, 180, 203, 205, 220, 226–228, 232, 235, 244, 246, 251, 252
IBM compatibility 41

ICI 58, 62, 64, 67, 68, 72, 95, 96, 127, 143, 145, 174, 177, 179, 215, 228, 264
ICL 40–42, 50, 51, 54, 65, 67, 69, 83, 85, 88, 95, 96, 160–162, 213, 216, 228, 230–233, 235, 242, 243
IEPG (Independent European Programme Group) 185
ILO (International Labour Office) 124, 264
Image processing 185
Immobilization of enzymes 12, 13
Immunex 241
Imperial Metal Industries 195
Import penetration 132, 133
Independent Business Units (IBUs) 84–86
Industriegewerkschaft Metall 107, 264
Inflation 2, 162
Inflexibility 2, 4, 26
Infrastructure 1, 25, 171, 257
Inmos 34, 35, 39, 161, 199, 240, 245
Innovation ix, xii, 5, 36, 66, 76, 119, 128, 130, 141, 146, 150, 151, 156, 164, 170, 196
Insurance 103
Intel Inc. 8, 34, 36, 84, 132, 238, 239
Intercom 236
Intersil 34
INSTIR 170
Institutions xii, 256
Integrated Circuits, *see* Semiconductors
Interest rates 3, 118
Intergraph 51
Internationalization 89, 156, 158, 164, 177, 197, 198, 206, 207, 245, 254
Intrapreneurs 75, 84, 85
Inventories 3, 22, 23, 25, 74, 76, 77, 92, 94, 118
Investments 1–4, 26, 62, 63, 76, 80, 104, 109–111, 117, 137, 167, 196, 204, 256
Ireland 136, 173, 190, 229
IRI 68, 69, 213, 228, 231
IRM viii
ISDN (Integrated Services Digital Network) 136, 137, 169
Isoglucose 145
Isuzu 79
Italable 195
Italtel 44–46, 190, 195, 204, 236, 237, 264

Italy 33, 41, 44–47, 49, 58, 119, 131, 136, 138, 139, 151, 158, 162–164, 170, 171, 173, 179, 186, 188, 190, 204, 229, 233
ITT 34, 39, 42, 45, 46, 72, 135, 161, 168, 169, 189, 190, 195, 203, 226–232

Jacobson, S. 110, 262
Jacquemin, A. 234, 264
Jaeger 80
Japan xii, 1, 13, 16, 48–50, 60, 76–79, 89, 92, 94, 105, 106, 108, 115, 130, 131–145, 147–151, 153–155, 157–159, 173, 178, 180, 183, 186, 188, 203, 255
Japanization xi, 81, 92, 94, 96
Jasanoff, S. 173, 174, 264
Jeandon, J. P. 99, 181, 264
Jethanandani, K. 262
Jeumont–Schneider 237, 243
Joint Ventures 35, 42, 47, 51, 84–87, 200, 244–245
Jones, B. 39
Jones, D. 260
Jürgens, U. 76–78, 109, 110, 113, 264, 266
Junne, G. 156, 234, 264
Just-in-time 77, 80, 81, 92, 93, 119

Kabelmetal 200
KabiGen 61, 242
Kabi Vitrum 241, 242
Kanban, *see* Just-in-time
Kaplinsky, R. 7, 10, 264, 265
Kariamo, K. (Nokia) 214
Kartte, W. 200
Kaske, K.-H. (Siemens) 214
Kawasaki 238, 245
Kenematsu Gosho 243
Kennedy–Round 3
Kenney, M. 260
Kern, H. 116, 265
Keynes, J. M. 1
Kleinknecht, A. 265
Klodt, H. 171, 264
Kloppenburg, J. 260
Know-how 3
Kokusai Denshin 235
Kollmorgen 241
Komatsu 127
Kontratieff, N. D. 1, 265

Kreye, O. 263
Kristiansen, B. 145, 265
Krupp 61, 65, 67, 68, 87, 95, 127, 187, 193
KUKA 47, 238
Kyocera 235, 243

Labarrère, C. 190, 265
Labour intensity 3, 94–97, 118
Labour market 114, 121
Labour process ix
Labour relations xi, 4, 151
Läpple, D. 94, 265
Landaburu, E. viii
Landes, D. 265
Large Scale Integrated Circuits (LSI) 33, 35
Lasers 15
Lead time 26, 50
Learning period 109, 110
Leo–Ferrosan 61
Levie, H. 122, 265
Lewis, C. 145, 265
Licences 49, 52, 234, 247
Liebhaberg, B. 215
Life-cycle 5, 28, 128, 129, 147, 196, 211
Life-time employment 114
Lilly, M. D. 260
Litronix 35
Little, Arthur D. 135, 137
Local Area Networks (LANs) 45, 83
Lockheed 51
Locksley, G. 91, 265
Logica 242
Logistics 21, 22, 54, 108, 118, 123
Long waves 1
Lorenz, G. 36, 265
Louisse, J. 185, 265
Low technology 73
Lucas 79
Lüthje, B. 206, 265
Luxemburg 173, 229
Luxury products 5
Lyonnaise des Eaux 214

Machine tool industry 47, 48, 76, 111, 139, 169–171
'Machino-facture' 75
McCorduck, P. 10
Mackenzie, I. 40, 265
Mackintosh, I. 34, 137, 265
Madelin 165

Magana, L. 214
Mainframe computers 8, 10, 21, 41, 42, 54, 84, 159, 168, 246
Maintenance xii, 44, 101, 112
Malerba, F. 33, 36, 131, 132, 265
Malsch, T. 76–78, 109, 110, 113, 264, 266
Malsot, J. 129, 266
Management xii, 23, 26, 74, 83, 85, 86, 116, 122, 131, 146, 147, 152–155, 185
Mannesmann AG 45, 65, 69, 72, 95, 195, 203, 228
Mantec GmbH 48
MAP (Manufacturing Automation Protocol) 80, 85
Marketing 56, 58, 89, 103, 104, 128, 129
Market shares 132, 133, 145
Market size 132, 135, 137
Markus, M. L. 152
Markusen, A. R. 266
Maroni 50
Mass production 4, 5, 26, 29, 31
Massachusetts General Hospital 57
Matra 32, 49, 51, 52, 55, 69, 72, 95, 160, 172, 192, 195, 196, 201, 203, 216, 228, 232, 236, 237, 239
Matra–Harris 39, 160, 187
Matsushita 34, 66, 92, 127, 178, 203, 238
Maturity 1, 105
Maucher, H. (Nestlé) 214
Max-Planck-Institut für Züchtungsfoschung 57, 174
Mazda 140
MBB (Messerschmitt–Bölkow–Blohm GmbH) 52
MCC 199
MCI 45, 203
McCorduck, P. 262
McDonnal Douglas 235
McKinsey 181, 266
McLean, M. 32, 33, 35, 187, 266
McMillan, C. 92, 266
Mechanization 74, 75
Mechatronics 71
Megachip 38, 91, 166, 167, 181, 184, 224
Membranes 15, 24
Memory chips 9, 35, 37, 38
Mensch, G. 1, 266

Mentor 52
Mercedes, *see* Daimler–Benz
Merchant production 31, 35
Mercury 72, 203, 204
Mergers 33, 38, 40, 45, 50, 160–162, 168, 198–200
Merkle, H. (Bosch) 214
Merrit, G. 215
Metze, M. 94
Meyer, A. de 154, 266
Michalet, C.-A, 266
Michelin 70
Microelectronics ix, xi, xiii, 6–11, 13–16, 20, 21, 24, 25, 30, 33, 57, 66, 71, 82, 90, 101, 105, 118, 121, 141, 142, 152, 159, 163, 164, 166–172, 177, 207, 212, 248
Microorganism 12, 13
Microprocessors 8–11, 16, 33, 35–37, 40, 53, 102
Microsoft 11, 55, 56, 84, 242
Microwave semiconductors 35
Middle East 61
Middle management 116, 255
Mietek 39
Miles Inc. 57, 143
Military 7, 32, 33, 35, 38, 83, 131, 135, 158, 159, 161, 166, 167, 171, 181, 185–188, 192–195, 201, 207
Miniaturization 8, 15, 23
Minicomputers 41
Minimum sales 36
Mining 56
MIT (Massachusetts Institute of Technology) 81, 92, 152
Mitel 199, 203, 235
MITI 149, 164, 266
Mitsubishi 34, 78, 127, 235
MMI 34
3M 62, 240
Mobile car telephone 136
Mobility 151, 153, 154
Moët–Hennessy 60
Molecular Diagnostics 57
Monetarism 2
Monitoring 24, 25
Monoclonal antibodies 13–14
Monod, J. 214
Monopolies and Mergers Commission (MMC) 199
Montedison 58, 69, 177, 179, 241
Moon, J. 166, 196, 268, 270

Moore, R. 122, 265
Moses, V. 59, 266
Mostek 34, 38, 72, 160, 199
Motorola 34, 36, 37, 132, 160, 238
MS-DOS 11
MTU 187, 200
Mullard (Philips) 161
Murray, F. 80, 81, 266

Naarden International 60
National champions xiii, 40, 43, 44, 135, 136, 168, 171, 186, 188, 191, 199, 200, 202, 257
National Coal Board 70
National Institutes of Health 57
National semiconductors 34, 132, 160, 239
Nationalization 42, 44, 175, 196, 197, 207
NEB (National Enterprise Board) 161, 174
NEC (Nippon Electric Company) 34, 42, 127, 132, 135, 235, 239, 242, 246
Nestlé 58, 60, 68, 144, 213, 216, 228, 241
Netherlands 38, 41, 44, 56, 60, 63, 93, 94, 107, 120, 136, 143, 147, 162–164, 166, 169, 170, 173–175, 178–180, 184–186, 188, 190, 229, 231
Networks 216, 218, 231, 247
Neumann, J. v. 10
New materials ix, xii, 14, 15, 22, 23, 25, 26, 31, 61–63, 68, 69, 71, 72, 114, 181, 185, 222, 240, 241, 248
Newly Industrializing Countries (NICs) 2, 3, 61, 125
Niche strategy 32, 33, 66, 187
Nicolin, C. (ASEA) 214
Nippon Steel 238
Nissan 77–79, 96, 127, 140, 178
Nixdorf 35, 41, 43, 45, 54, 65, 67, 69, 70, 95, 127, 160, 194, 195, 203, 213, 228, 216, 232, 243
NMB (Ned. Middenstandsbank NV) 94, 266
Noble, D. 17, 266
Nokia 214, 236
Nomura Research Institute 132
Nordisk 241
Norsk Data 243

Norsk Hydro 214
Northern Telecom 44, 135, 190, 235, 236
Norway 52, 115, 162, 190, 229
Novo 56, 143, 177, 242
NTT 203
Nuclear power 7, 159, 182, 183
Numerically controlled machines 7, 20, 26, 79, 100, 170, 171, 219

Océ–van der Grinten 93
ODETTE (Organisation des Données Echangées par Télétransmission en Europa) 80
OECD 1, 2, 4, 9, 20, 53–55, 79, 91, 102, 106–108, 115, 133, 134, 138–140, 150, 151, 162, 163, 177, 182, 183, 186, 188, 234, 266, 267
Office equipment 43
Office of Technology Assessment (OTA) 16, 51, 52, 55, 58, 81, 102, 103, 107, 108, 132, 142–144, 175, 182
Office-of-the-future x, 17–19, 43, 66, 67, 83, 102, 219
Off-shore 71
OGE 263
Ohmae, K. 50, 92, 125, 267
Oil industry 56, 71–73, 126
Oki 34
OKK 49
Olivetti 30, 39, 41–43, 45, 49, 54, 58, 65, 67, 69, 72, 81, 95, 160, 179, 195, 204, 205, 213–216, 220, 228, 230–233, 236, 238, 239, 242, 243, 246, 258
Olympia 42
Opel AG 85, 86
Optical instruments 6
Optical fibres 9, 15, 23, 218
Optronics 61, 218
Osai 49
OTA, *see* Office of Technology Assessment
Overbeek, H. 155
Overproduction 3, 32

PABX (Private Business Exchanges) 45, 83, 136, 203
Panneburg Commission 164
Pannenborg, A. E. 213, 227, 267
Pantoufflage 197

Paribas 205
Part-time work 115
Patel, P. 150–152, 154, 267
Patent litigation 217
Pavitt, K. 150–152, 154, 267
Pechiney 240
Pelissolo, J.-C. 175, 267
Penicillin 12
Pennen, T. v.d. 265
Pentagon, *see* US Department of Defense
Personal Computers 5, 10, 11, 21, 42, 84
Pesticides 57
Petit, P. 99, 260
Petre, P. 267
Petrochemical industry 1, 22, 58, 71, 177, 230
Petrofina 70
Peugeot (PSA) 31, 47, 51, 65, 67, 68, 122, 140, 178, 232, 240
Pharmaceutical industry 7, 14, 56–58, 60, 61, 66, 77–79, 87, 95, 143–146, 177, 230
Pharmacia 61, 242
Philips 12, 19, 32, 34–36, 38–42, 44, 46, 48, 54, 64, 66–68, 72, 83, 87, 90, 92–95, 102, 113, 120, 127, 132, 160, 161, 166, 167, 169, 178, 180, 182, 185, 190, 194, 200, 201, 203, 213–216, 224, 227, 228, 230–233, 235, 236, 238–240, 248, 249, 258, 267
Pierrel 242
Pilkington Bros. 214
Pino, A. 103, 261
Piore, M. J. 267
Piper, A. 204, 267
Pipework 52
Pirelli 214, 216, 240
PKI (Philips Kommunikations Industrie AG) 83
Plan Calcul 159
Plastics 6, 15, 25, 32, 63
Plessey 32, 34, 35, 42, 46, 135, 167, 179, 181, 187, 189, 190, 192, 203, 213–216, 228, 232, 235, 236, 240, 244
PMI 34
Polarization 115
Politics xii, 257
Pollution 2, 4, 19, 24, 25, 118

Polygram 72
Polymers 59
Porat, M. U. 267
Porsche 51
Portugal 190, 229, 231
Positional goods 5
Post-War boom x, 1, 2, 6, 23, 26, 29
Power 254, 255
Prahalad, C. 234, 246, 262
Pre-competitive research 31, 257
Preservatives 60
President's Commission on Industrial Competitiveness 148, 267
Preussag 70
Price/performance ratio 21, 110, 135
Private Business Exchanges, *see* PABX
Privatization 39, 196, 197, 202–204
Process innovation 17, 98–108
Product cycle 129
Product innovation 14–17, 98–106, 211
Productivity 2, 6, 7, 81, 103, 104, 106, 108, 187
Professional equipment 5
Profits 5, 29, 33, 58, 62, 66, 73, 101, 253
Protectionism 106, 135, 136, 158, 159, 201, 218
Protein engineering 11, 50
PSA, *see* Peugeot
Puteen 58
PTTs 9, 44, 45, 136, 137, 169, 189, 191, 202, 203
Public ownership 196
Public procurement 7, 40, 92, 100, 157, 159, 164, 177, 187–189, 192–195, 197, 202, 207
Public switching exchanges 45, 46, 53, 83, 168, 184, 189, 190, 200, 202
Purchasing power 106
Pyramid Technology Corp. 243

Qualifications 97, 112–116, 121, 150
Quality of production 25, 77, 80, 81, 92, 93, 108
Quandt group 47

Racal 50, 52, 243
Race viii, 38, 40, 45, 112, 128, 138, 158–177, 183, 208, 250

RACE (Research and Development in Advanced Communications Technology in Europe) 136, 168, 223–232, 246
Radar systems 10, 30
Radio 105
Rank Xerox 93, 94
Ransburg 237
Rathenau Commission 164
Rationalization 76, 85, 97, 98, 103, 109, 110, 115, 116, 122, 125, 165, 171
Raw materials 2, 4, 19, 22, 23, 25, 26
Raytheon 34
RCA 34, 238
Reagan, R. 199
Recession ix, 1, 29, 98
Recombinant DNA 13, 14
Reflet, J.-C. 39
Regional impact 116–120
Regulation 3, 4, 100, 145, 146, 197–208, 254
Reich, R. B. 13, 267
Reliability 8
Research and Development (R&D) ix, 6, 16, 28, 31, 36, 37, 39, 42, 43, 46, 49, 63–66, 70, 71, 88–90, 104, 112, 118, 126–129, 131, 135, 137, 145, 148, 150, 154, 163, 164, 168, 175, 178–183, 186, 210, 227, 256
Renault 20, 31, 47, 49, 50, 51, 65, 67, 68, 77–80, 87, 89–91, 95, 109, 113, 140, 172, 178,181, 194, 216, 228, 237, 240
Reorganization 82–94
Restructuring xii, 1, 66, 73, 91, 97, 117, 164, 171, 196
Restructuring race xiii, 156, 198, 199, 256
Rhodes, M. 181, 182, 267
Rhône–Poulenc 64, 68, 72, 89, 95, 143, 175, 177, 178, 195, 228
Richardson, J. J. 166, 191, 196, 267, 268, 270
Richman, L. S. 268
Rifa AB 37
Risk taking 142, 147
Rivalry xiii, 156
Roberts, D. H. (GEC) 214, 227
Robotronics 172
Robots 8, 9, 14, 16, 20, 29–31, 46–50, 68, 69, 71, 76, 77, 79, 100, 102–104, 110, 138–141, 165, 170, 171, 180, 200, 219, 237, 238, 244

Rockwell 34, 135, 229
Rodenburg, N. 20, 268
Rönning, G. 98, 263
Rolls Royce 21
Rolm 45, 235, 244
Roobeek, A. viii, 6, 19, 61, 145, 146, 268
Roos, D. 260
Roscam Abbing, M. viii
Rothwell, R. 164, 183, 187, 268
Round Table of Industrialists 213–215, 222, 225
Rowland, T. 32, 33, 35, 187, 266
Roussel Uclaf 175, 241
Ruhrkohle AG 70
Ruivenkamp, G. viii, 59, 268

Saab–Scania 39, 51, 72, 81, 168, 179
Sable, Ch. F. 5, 267, 268
Saint-Gobain 65, 68, 87, 95, 160, 195, 196, 214, 226, 228, 239, 240
Salisch, H. 115, 268
Sandoz 179, 241
Sanger, D. 84
Sanofi 59
Sankyo Seiki 49, 237
Santilly 109
Sargeant, K. 23, 268
Sasson, A. 174, 268
Satellite Business Systems 45
Satellites 9, 45
Saunders, C. 171, 264
Saussay, P. 261
Savary, J. 196, 268
Saving rate 148
SCEMI 49
Scharpf, F. W. 268
Schering 174
Schering–Plough 58, 242
Schlumberger 34, 35, 51
Schmidheiny, S. (Anova) 214
Schmitz, H. 268
Schumann, M. 116, 265
Schumpeter, J. A. 1, 2, 268
Schutte, G. 223, 268
Scicon 55
Scott, R. 58, 269
SCP, *see* Single-Cell Proteins
SCS–ATES 69
SDI (Strategic Defence Initiative) 223, 247
Secmai 52

Second sourcing 33, 36, 37, 55, 70, 246
Seeds 60
Segmentation 116, 254
Seiko 72
SEL (Standard Elektrik Lorenz) 72, 189, 200, 203
Selenia 51, 87
Self-diagnosis 22, 117
Self-repairing machinery 22, 117
Semiconductors 8, 9, 11, 14, 24, 36–40, 61, 68, 69, 105, 113, 119, 130–134, 137, 159–161, 166, 187, 200, 238–240
Semiconductor Industry Association 130, 269
Senessons 61
Sensors 13, 47, 103, 110, 141
Serono 242
Servan-Schreiber, J. J. 126, 269
Services 74, 97, 101, 103, 115, 118, 136, 137
Sescosem 34
SGS 33, 36, 38
SGS–ATES 33, 34, 38
Shaiken, H. 122, 152, 153, 269
Shakeout 138
Shape Data (UK) 52
Sharp Inc. 34
Sharp, M. 12, 57, 143, 152, 155, 175, 262, 264, 269
Sheehy, P. 214
Shelf life 60
Shell 59, 64, 67, 68, 71, 90, 95, 96, 127, 174, 178, 179, 215, 241, 269
Shiftwork 3, 111, 120, 123
Sibeka SA 214
Siemens AG 32, 34–36, 38–42, 44–49, 51–55, 64, 66, 68, 72, 81, 83, 85, 91, 95, 112–115, 127, 132, 135, 136, 141, 160, 166–169, 178, 181, 182, 187, 189, 190, 192, 200, 201, 203, 205, 206, 213–216, 224, 228, 230–233, 236, 237, 239, 240, 242, 243, 245, 246, 249, 256, 269
Signetics 34, 35
Silicon Glen 15
Silicon Valley 15, 24, 118, 167
Siliconix 34
Silicium 61
Single-Cell Proteins (SCP) 57–59, 174, 217
SIP (STET) 204
Sitter 269

Size of companies 126, 257
Skills, *see* Qualifications
Small and medium sized enterprises ix, 80, 94, 97, 98, 100, 107, 108, 116, 169–171, 180, 182–185, 188, 207, 208, 226, 227, 233, 257, 259
Small countries 167, 168, 172, 176, 180, 183, 191, 198, 201, 208, 225, 231, 259
Smokestack industries 24, 25
SNA (Systems Network Architecture) 84
SNIAS 182
Sobel, R. 159, 269
Social context xii, 97–124, 146, 155, 255
Socialist countries 125
Sodeteg 59
Software 8, 11, 14, 16, 30, 43, 51–56, 68, 69, 71, 84, 86, 87, 89, 108, 141, 142, 168, 242, 243
Solana, L. 214
Solvay 177
Sony 127
Sormel 172
South-East Asia 119, 120
SPAG (Standard Promotion and Application Group) 215, 216, 230
Spain 136, 190, 229
Specialization x, xii, 31–33, 47, 63, 67, 71, 73, 74, 82, 123, 254
Spethmann, D. (Thyssen) 214
Sperry 40, 242
Spinks, A. 174
Spin-off 33, 55
Sprague 34
Sprangers, C. 55
Squibb & Sons 242
Stagnation 1, 66, 111
Standardization 36, 46, 56, 80, 87, 90, 98, 127, 135, 141, 168, 206, 216, 220, 245, 250, 257
Stanford University 12
Stansaab 168
Starch 58
'Star Wars' 17
Start-up firms 35, 39, 52, 56, 88, 131, 143, 147–149, 181
Stauffer 72
STC–ICL 41, 42, 46, 69, 83, 161, 167, 193, 199, 203, 213, 228
Steel industry 62, 63, 90, 164
Steenbergen 205, 206, 261

Steinbrinck, R. 94, 269
Stevens, A. 189, 197, 261
Stern, J. (Bull) 42
STET group 33, 38, 44, 49, 51, 65,
     67–69, 87, 179, 192, 195, 197, 204,
     213, 216, 228, 237–240, 242, 245,
     246
Stocks, *see* Inventories
Stopford, J. 195, 269
STR (ITT) 168
Stratus Comp. 243
Strikes 81, 87
Stromberg–Carlsson 46, 203
Sweden 41, 44, 51, 52, 61, 103, 104,
     107, 110, 115, 130, 136, 138, 139,
     144, 147, 148, 151, 155, 162, 163,
     167, 168, 170, 171, 179, 180, 183,
     184, 186, 188, 189, 190, 208, 229,
     256
Switching equipment 9
Switzerland 58, 130, 136, 142–144,
     147, 151, 155, 162, 163, 168, 179,
     180, 186, 190, 208, 229, 256
Subcontractors xi, xii, 74, 76, 77,
     79–82, 85, 89, 92–94, 96, 97, 107,
     108, 112, 118–120, 123, 184, 188,
     205
Submicron technology 36, 38
Subsidies 156–158, 162, 163, 166–169,
     173, 175, 182–186, 191, 192–195,
     197–199, 207, 223, 224, 257
Sumitomo Electric 127, 240
Suppliers, *see* Subcontractors
Svanholm, P. J. 214
Swan, D. 212, 269
Sycpro 87
Synchrotron project 160
Synertek 34
Synthetic materials 61–63
System X 46, 189
Systems houses 96

Tactile sensitivity 9
Takeda Pharmaceutical 241
Takeovers 30
Talbot 122
Tariffs 257
Tastes 5
Taus, J. 214
Taxes 63, 157
Taylorism 75, 89
Technology gap 155

Technology parks 147
Technology push 40
Technology transfers 129, 153
Technology webs 6
Tekade 190
Telecommunications 8, 9, 14–16, 21,
     23, 26, 32, 37, 42–46, 52, 68, 69,
     83, 90, 91, 104, 115, 117, 118,
     133–138, 159, 166, 168, 169, 181,
     188–195, 198–203, 235–237, 244,
     245, 248, 254
Teleconferencing 89, 90
Telecopy machines 9
Telefonica 39, 214, 243
Telefonbau und Normalzeit (Telenorma)
     45, 72
Telefunken 72
Telenokia 190
Telespazio (STET) 195
Telettra (Fiat) 46, 190, 235
Televerkeet 189, 191
Television sets 8, 9, 19
Teli 189
Tertiary recovery of oil 59
Texas Instruments (TI) 34, 39, 132,
     159, 238
Thermoplastics 62
Third World 3, 106, 117, 120, 123,
     124, 134, 248
Thomson 32, 34, 36–39, 45, 46, 49, 55,
     65–69, 72, 83, 95, 113, 119, 135,
     160, 167, 178, 181, 182, 190, 192,
     196, 197, 199, 200, 203, 213–216,
     228, 230–233, 236, 237, 239, 240,
     246
Thorn–EMI 34, 39, 55, 69, 71, 72, 83,
     95, 161, 190, 193, 199, 203, 228,
     230, 235, 237, 240, 269
Thorn–Ericsson 195
Threshold Technology 35
Thyssen 70, 214
TIE 235
Tillesse, P. de (Pirelli) 214
Tindemans 168
Tissue culture 60, 143
Thoughtware 11
Toffler, A. 204
Toho Rayon 241
Tokyo Round 3
Tolsma, H. 55, 120
Toshiba 34, 39, 66, 132, 178, 236, 237,
     239, 245, 249

Toshiba Seiki 49
Toxic substances 24
Toyobo 240
Toyo Kogyo 78, 79
Toyo Soda 241
Toyota 77–79, 96, 140, 178
Trade barriers 77, 127, 158, 201
Trade unions 82, 85, 86, 109, 111, 112, 120–123, 151, 152, 165, 203, 205, 255
Trading activities 74, 96, 97
Traditional sectors xii, 7, 218
Training 110, 112–116, 150, 151, 170, 256
Trallfa 50
Transfer of technology xii, 41, 43, 245, 247
Transistors 7
Transport 3, 21, 23–25, 75
Triad Power 50
Trilogy 236, 240
Triumph–Adler 30, 42, 43, 61, 72
TRW 34
Truel, J. L. 37, 269
Tsurumi, Y. 77, 269
Tulder, R. v. 29, 234, 264, 269
Turkey 229
Turner, L. 269
Tymshare 236
Typewriters 20, 42, 43

UCSL 55
Unemployment 1, 162
UNICE (Union des Industries de la Communauté Européenne) 211
Unidata 41, 160, 167
Unilever 55, 58, 60, 64, 68, 72, 86, 90, 91, 95, 143, 144, 178, 179, 215
United Kingdom 16, 32, 33, 39–41, 43, 44, 48–50, 52, 54, 55, 77, 92, 94, 105, 114, 115, 119, 122, 130, 131, 136, 138, 139, 142–144, 148, 151, 152, 161–164, 166, 170–175, 179, 180, 183, 186–189, 190, 199, 201, 203, 204, 212, 229, 256
United Nations 4, 269
UNCTC (United Nations Centre on Transnational Corporations) 91, 269
United States of America ix, xii, 1, 6, 7, 20, 33, 39–41, 43, 48, 50–52, 56–58, 76–78, 84, 89, 91, 94, 100, 101, 103–106, 115, 118, 120, 124, 130–151, 153–155, 157, 158, 167, 168, 171, 173, 174, 178, 180, 182, 186–188, 199, 203, 204, 210
US Air Force 17
US Department of Commerce 6, 40, 53, 54, 130, 133, 159, 270
US Department of Defense 7, 16, 57, 135, 159, 187
US Office of Labor Statistics 100
United Technologies Inc. 38, 66, 72, 160, 178, 199, 235
Unitika 240, 245
Universities 57, 58, 88, 89, 92, 114, 143, 144, 146–149, 150, 173, 174, 176
UNIX 11, 54
Urbanization 29
Uri, P. 126, 269
USSR 16

Vaccines 143
Valvo, *see* Philips
Value Added Networks (VANs) 45, 83, 202–205
Vauxhall 122
VDI–Technologiezentrum 105
Venture Capital 39, 143, 148, 149
Verschure, J. 115
Vertical integration 33, 34, 79, 81
Very large scale integration (VLSI) 8, 36, 38, 52, 167, 222
Vlek, R. 265
Vision systems 9, 76, 110, 172
Volkswagen (VW) 30, 31, 47, 51, 61, 65, 67, 68, 72, 77–79, 81, 95, 110, 127, 140, 178, 194, 228, 230, 240
Volvo 31, 47, 51, 61, 65–68, 78, 79, 81, 93, 95, 179, 194, 195, 214, 216, 228, 240
Vyoril 177

Wages 2, 3, 5, 95–97, 118
Wagner Commission 164
Walkman 105
Wang 243
Washing machines 20
Waste 23–25
Waterford Glass 214
Webber, D. 166, 182, 196, 206, 268, 270

Wellcome 143, 177
Welzk, S. 149, 270
Western Electric (AT&T) 44, 203, 204
West Germany 38, 40, 41, 43–45, 47, 49, 52, 55, 57, 63, 76, 78, 82, 83, 103, 105, 109, 110, 111, 113, 115, 119, 120, 130, 136, 138–140, 142–144, 147, 150–153, 155, 158–160, 162–166, 169–171, 173–176, 178, 180, 182–184, 186–189, 190, 200, 203, 205, 208, 229, 256
Westinghouse 16, 127, 238, 245
White collar employment 24
Williamson, O. E. 90, 270
Wiltgen, P. viii, 43, 60, 90, 179, 270
Wit, A. de 54, 270

Womack, J. 260
Women 115, 116
Working conditions 165
World economy 106
Word processors 20, 102, 110

Xerox 236

Yankee Group 151, 270
Yaskawa 49, 237, 245
Yorke, K. 234, 263
Yoxen, E. J. 174, 270

Zarader, R. 99, 181, 264
Zegveld, W. 183, 187, 268
Zilog 34, 36, 239
Zimmermann, J.-B. 262